L. Dallas Tuck

Principles of Solution and Solubility

UNDERGRADUATE CHEMISTRY

A Series of Textbooks

edited by
J. J. Lagowski
Department of Chemistry
The University of Texas at Austin

Volume 1: Modern Inorganic Chemistry, *J. J. Lagowski*

Volume 2: Modern Chemical Analysis and Instrumentation, *Harold F. Walton and Jorge Reyes*

Volume 3: Problems in Chemistry, Revised and Expanded, *Henry O. Daley, Jr. and Robert F. O'Malley*

Volume 4: Principles of Colloid and Surface Chemistry, *Paul C. Hiemenz*

Volume 5: Principles of Solution and Solubility, *Kōzō Shinoda, translated in collaboration with Paul Becher*

Principles of Solution and Solubility

KŌZŌ SHINODA
Yokohama Nation University
Yokohama, Japan

Translated in Collaboration With

PAUL BECHER
ICI Americas Inc.
Wilmington, Delaware

MARCEL DEKKER, INC. New York and Basel

Library of Congress Cataloging in Publication Data

Shinoda, Kozo, [Date]
Principles of solution and solubility.

(Undergraduate chemistry ; v. 5)
Translation of Yoeki to yokaido.
Includes index.
1. Solubility. 2. Solution (Chemistry)
I. Title. [DNLM: 1. Solubility. 2. Solutions. QD543 S556y]
QD543.S5513 541'.34 78-13385
ISBN 0-8247-6717-9

MARCEL DEKKER, INC.
270 Madison Avenue, New York, New York 10016

Current printing(last digit):
10 9 8 7 6 5 4 3 2 1

PRINTED IN THE UNITED STATES OF AMERICA

FOREWORD

Dr. Kōzō Shinoda was one of my most skillful collaborators from 1957 to 1965 in the development of regular solution theory. We were co-authors of nine papers on aspects of the theory.

His subsequent career as a writer and teacher in this field, and his contributions to knowledge of complex systems involving surfactants, have afforded me much satisfaction.

Joel H. Hildebrand

PREFACE

Chemistry was once defined as the study of solutions because most reactions occur in liquids.

J. W. McBain

So expressed, the study of solutions is indispensable for most chemists and chemical engineers. There are several important monographs on solutions, each embodying a rather different point of view. These books are fairly massive, deal with randomly mixed solutions, yet do not contain material on many systems of practical importance, i.e., these are more or less books for specialists. Nowadays, however, science is specialized into so many fields that it is almost impossible for chemists or other scientists to master many of them.

Accordingly, a book on Solution and Solubility, realistic, brief, and suitable for efficient study is desirable for chemists, in order to further our own research in our respective fields. Such a book should contain chapters on solubility in water, solutions of polar-nonpolar molecules, surface-active substances, water-soluble polymers, solubilized solutions, and so on, which are important for practical problems and of importance to most research chemists. The most striking feature of these solutions may be the dissolution due to the orientation, arrangement, and structure formation of molecules, which are otherwise practically insoluble by random mixing only.

This book has been written in response to these demands, to give a coherent treatment of the theory of solutions, and to stimulate further research by inviting attention to the present limitations in our knowledge of the subject.

Present-day life is too nervous, busy, and we are assailed on all sides by information. Such surroundings are, I believe, a disadvantage (which older generations avoided) in examining the main issue of entire phenomena correctly, and in proceeding with research in the correct direction. The author intends that the reader may learn not only about solutions, but may be inspired to originality and creativity through the study of this book.

Constant fresh interest in one's research or profession may be the greatest happiness for the research scientist. Professor J. H. Hildebrand of the University of California is a distinguished scientist, who established the regular solution theory. He is still continuing his research with fresh interest at the age of 96. I would like to quote the following from his writings (*American Scientist*, March 1963, pp. 4-5).

"I have for years had able post-doctoral research collaborators. If I had been really retired in 1952 (70 years), instead of pseudo-retired, I would have been deprived of the ten most productive, influential, and exciting years of my life.

"...during my long life I have had great fun devising searching questions to put to Mother Nature and interpreting her often surprising answers. And I am still doing it zestfully at the age of eighty-one...."

The translation of my manuscript was performed with the collaboration of Dr. P. Becher of ICI Americas Inc. Dr. Masuo Sugai of Kao Soap Co., Ltd. made an early contribution. This book would not have appeared without their cooperation. I sincerely thank them for their great effort.

Solution and Solubility (in Japanese) was published in 1966, revised edition in 1974, by Maruzen Co., Ltd., Tokyo. The author thanks the publishers who kindly gave their permission for the English version of this book.

Kōzō Shinoda

TRANSLATOR'S PREFACE

When, almost ten years ago, Professor Shinoda, Masuo Sugai, and I undertook to translate the first edition of *Solutions and Solubility* into English, we lightheartedly assured the publisher that the manuscript would be complete in a year--at most, eighteen months. It is a tribute, I think, to the value of Professor Shinoda's book that the publisher has stayed with us, encouraged us, needled us, and kept us at our task. Dr. M. Sugai, because of the pressure of other duties, unfortunately had to bow out at an early date.

The book, as it now stands, is closer to the second edition of *Solutions and Solubility* than to the first, which was our initial target. However, Professor Shinoda has used the opportunity to rethink, rewrite, and to accept generously the thoughts of his collaborator, so that this volume is, in effect, a new book. We have recognized this fact by changing the title slightly.

I have referred to the charitable acceptance of the long delay in the completion of this book by the publishers. We would also like to acknowledge our debt to Professor Stig Friberg who critically reviewed the major part of the manuscript and made invaluable suggestions for improvement. I must express my own appreciation to my secretary, Miss Di Leonardo, who has cheerfully typed uncounted drafts of the manuscript.

When a long effort is complete, there is a sense of loss. Appropriately, my feeling is reflected in a *haiku* by Shiki:

The others go home,
With the fireworks over,
how dark it's become!

Paul Becher

CONTENTS

Principles of Solution and Solubility

chapter one

INTRODUCTORY SOLUTION THERMODYNAMICS

1.1 THE SCIENTIFIC MIND

Our everyday experiences with solutions are important experimental facts. Wisdom, judgment, and the desire to take into account these facts in our understanding of the properties of solutions or in the interpretation of the state of solution, dissolution mechanism, and so on, are the most important ingredients. Without these ingredients the total and coherent understanding of various phenomena observed in solutions of widely different types would seem impossible.

Let us consider some examples:

1. We observe, in fertilizing our garden, that dissolving ammonium chloride or ammonium sulfate in water causes it to cool down. If we observe this phenomenon with a scientific mind, we conclude that the process of solution of ammonium chloride is endothermic, i.e., that heat is absorbed from the surroundings and the system remains at constant temperature. In other words, the enthalpy increases because of the mixing. What, then, is the driving force for dissolution? We know from thermodynamics that the Gibbs free energy of the system is decreased when any reaction occurs at constant temperature and pressure. At constant temperature, the differences between the thermodynamic functions before and after mixing are given by

$$\Delta G = \Delta H - T\Delta S \qquad (1.1)$$

As the mixing is endothermic, the enthalpy of the system increases, i.e., $\Delta H > 0$, and the Gibbs free energy always decreases, i.e., $\Delta G < 0$. Hence, we observe that the entropy of the system increases owing to the dissolution of salts. In other words,

$$\Delta S > \frac{\Delta H}{T} \qquad \text{or} \qquad \Delta S = \frac{\Delta H - \Delta G}{T}$$

Why does the entropy of the system increase on dissolution? The entropy of the respective components increases because of the dilution, just as in the entropy of mixing of gases. However, if the process is fairly endothermic, a nonelectrolyte solute does not readily dissolve. The solubility will be very small, since the available entropy is used to compensate for the enthalpy increase.

On the other hand, we know that an electrolyte dissociates in water, and that the electrical conductivity increases; thus we observe that dissociation into ions results in a large increase in entropy, just as the dissociation of N_2O_4 into $2NO_2$ results in a large increase in the entropy of the system (mostly translational entropy, because of the increase in number of particles).

2. Recrystallization is a classical and powerful method of purification. For example, we observe the efficient purification of NaCl and $CuSO_4 \cdot 5H_2O$ by crystallization from water, or of organic crystals from organic solvents. From these well-known facts we should also note that mixed crystals rarely form when the sizes of molecules or ions are different. Therefore, crystallization or zone melting is a handy and powerful method of purification. At the same time, we note that molecules of differing sizes mix well in solution, and this is a characteristic of the solution process. If the sizes of the molecules are the determining factors in the process of solidification, we may expect that some pairs of molecules whose size and shape are appropriate will, in fact, form mixed crystals or clathrate compounds.

3. Analogy is also very important for the understanding of solution phenomena. For example, we can regard the vapor pressure as the solubility in a vacuum, i.e., in a fluid of zero

intermolecular force. Hence, the boiling point corresponds to a solubility value of a particular substance. We can evaluate the cohesive energy density (see Sec. 4.4) from the boiling point and the molar volume of a substance. The square root of this quantity is called the solubility parameter and is used to evaluate the energy of mixing and solubility in regular solution theory. That is, the solubility parameter is evaluated from the boiling point "solubility" of a substance into vacuum. We can draw similar inferences from the temperature dependence of the liquid and equilibrium vapor densities, as shown in Fig. 1.1 [1].

The liquid density and the equilibrium vapor density are related by the so-called rectilinear diameter rule,

$$\frac{d_\ell + d_g}{2} = d_c + a\left(1 - \frac{T}{T_c}\right) \tag{1.2}$$

where d_ℓ and d_g are the liquid and vapor densities, respectively, T is the absolute temperature, T_c is the critical temperature,

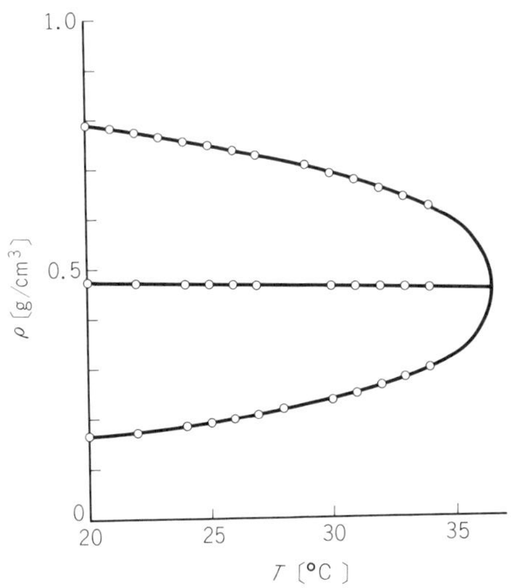

Fig. 1.1 The orthobaric densities of nitrous oxide [1]. The straight line through the mean densities is given by Eq. (1.2), with d_c = 0.452 and a = 0.00132.

d_c is the critical density, and a is a constant which is usually small in comparison with d_c. This law was first stated by Cailletet and Mathias in 1886.

In effect, $d_\ell + d_g$ is almost constant with temperature. This means that the total volume of holes in the liquid and the total volume (pressure) of molecules in the gaseous phase are nearly constant with temperature. The result is a critical density d_c roughly equal to half the density of the liquid at low temperatures. In fact, we often find that the volume or weight fraction of the respective liquids at the critical point of liquid-liquid mixture is nearly 0.5. This phenomenon is illustrated in Figs. 1.1 and 5.7.

Thus, daily experiences and analogies among universal facts aid us in developing a complete understanding of solutions, as well as creativity.

In this book we shall consider the following points:

1. Why a solute will or will not dissolve in a particular solvent.
2. Why similar solvents dissolve solutes to the same extent, yet themselves show different solubility in a third solvent.
3. Why there is a large difference in the solubility of isomers of different melting points.
4. How to increase the solubility of gases, liquids, and solids.
5. The mechanism of the dissolution of polymers, of surface-active substances as micelles, the mechanism of solubilization by micelles, salting-out phenomena in soaps, etc.

1.2 CORRESPONDING STATES; REDUCED EQUATIONS OF STATE

The van der Waals equation of state for 1 mol of a fluid is given as

$$\left(P + \frac{a}{V^2}\right)(V - b) = RT \tag{1.3}$$

where P is the pressure, V is the molal volume, T is the absolute temperature, R is the gas constant, and a and b are constants characteristic of the fluid. This equation is obtained by taking into account the effect of intermolecular forces on the equation of state for an ideal gas. The constant a is the correction factor for attractive forces, and b that for repulsive forces.

At the highest temperatures, the fluid is a nearly perfect gas, and the curve approaches a regular hyperbola. As the temperature is lowered, however, the P-V curve shows zero slope, i.e., $(\partial P/\partial V)_T = 0$. This is the critical point. Below this temperature two phases coexist, or in other words, this equation explains critical phenomena. Therefore, the van der Waals equation also describes the liquid state. Although the agreement with actual experimental curves in the dense fluid is only qualitative, the equation is important in explaining condensation, critical phenomena, and so on.

As the pressure is a function of V^3, V yields roots at moderate temperature and pressure, where

$$\left(\frac{\partial P}{\partial V}\right)_T = 0, \qquad \left(\frac{\partial^2 P}{\partial V^2}\right)_T = 0 \tag{1.4}$$

holds. From Eqs. (1.3) and (1.4) we obtain the relations among the critical temperature, pressure, volume, and the constants a and b.

$$a = 3P_c V_c^2, \qquad b = \frac{V_c}{3}, \qquad RT_c = \frac{8}{3} P_c V_c \tag{1.5}$$

Introducing these relations into Eq. (1.3), we obtain

$$\left(\frac{P}{P_c} + 3\frac{V_c^2}{V^2}\right)\left(3\frac{V}{V_c} - 1\right) = 8\frac{T}{T_c} \tag{1.6}$$

Equation (1.6) is called the *reduced equation of state*. Thus, all substances would have the same equation of state when expressed in terms of the reduced variables $T_r = T/T_c$, $P_r = P/P_c$, $V_r = V/V_c$, which are the ratios of the actual variables to the values of the critical temperature, pressure, and volume, respectively. This is called the principle of corresponding states. The critical point is clearly a corresponding state.

In Eq. (1.3), a and b are specific constants, so that equations

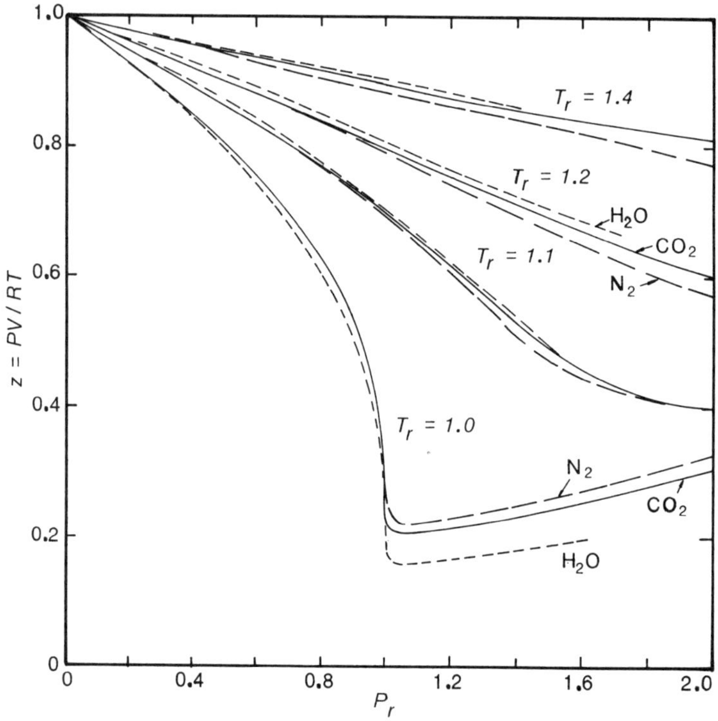

Fig. 1.2 A test of the hypothesis of corresponding states. Isotherms of N_2 (long dashed line), H_2O (short dashed line), and CO_2 (solid line). From *Thermodynamics*, by G. N. Lewis and M. Randall, 2d ed. revised by K. S. Pitzer and L. Brewer. Copyright 1961. Used with permission of McGraw-Hill Book Company.

of state are different for different substances. However, Eq. (1.6) is a universal equation of state, in that any reduced quantity for all substances is the same (as a first approximation) at the same values of the reduced variables. One example is shown in Fig. 1.2. This means that properties derived from the equation of state of a newly synthesized substance can be estimated from the properties of well-characterized (possibly similar) substances if the critical constants of the new substance are known.

The critical temperatures, pressures, and specific weights of a number of substances are summarized in Table 1.1. Critical

Table 1.1 Critical Temperature, Critical Pressure, and Specific Gravity (ρ_c) at Critical Point

Substances	T_c (°C)	P_c (atm)	ρ_c (g/cm^3)
CH_3COOH	321.6	57.2	0.351
$(CH_3)_2CO$	235.0	47	0.268
CH_3CN	274.7	47.7	0.240
$C_6H_5NH_2$	426	52.4	--
C_6H_6	288.5	47.7	0.304
C_6H_5Br	397	44.6	0.486
C_4H_9OH	287	48.4	--
CS_2	273	76	--
CCl_4	283.1	45.0	0.558
C_6H_5Cl	359	44.6	0.365
$CHCl_3$	263	--	0.516
cyclo-C_6H_{12}	281.0	40.4	0.270
C_2H_6	32.1	48.8	0.21
$CH_3COOC_2H_5$	250.1	37.8	0.308
C_2H_5OH	243.1	63.1	0.275
C_2H_5Cl	187.2	52	0.33
$(C_2H_5)_2O$	193.8	35.5	0.262
C_2H_4	9.7	50.9	0.22
n-C_7H_{16}	266.8	26.8	0.234
n-C_7H_{14}	234.8	29.5	0.234
CH_4	-82.5	45.8	0.162
$SnCl_4$	318.7	37.0	0.742
$C_6H_5CH_3$	320.6	41.6	0.292
NH_3	132.4	111.5	0.235
H_2O	374.0	217.7	0.4
Xe	16.6	58.2	1.155
Ar	-122	48	0.531
H_2	-239.9	12.8	0.031
He	-267.9	2.26	0.069

temperatures range from 5 K for helium to as-yet unexplored high temperatures. On the other hand, critical pressures for most substances are around 50 atm (40 to 60 atm). As the critical pressures are nearly constant for many substances, substances are at the same reduced pressure, ≈ 1/50, at their boiling point.

Thus, such reduced properties as $\Delta H_b^v/RT_b$, T_b/T_c, V_b/V_c at their boiling points are assumed to be nearly constant (two phases coexist at the boiling point, so that the state is fixed if one variable is determined). The constancy of the entropy of vaporization, $\Delta S_b^v = \Delta H_b^v/T_b \approx 21$ cal/deg mol, is known as Trouton's rule. The Guldberg rule states $T_r = T_b/T_c \approx 2/3$, and Walden's rule is $V_r = V_b/V_c \approx 1/3$.

Entropy of vaporization increases slowly with temperature, however. Hildebrand [2] described a substitute, known as the Hildebrand rule, which states that entropies of vaporization of unassociated liquids are equal, not at their boiling points, but at temperatures at which their vapors occupy equal volumes.

1.3 INTERMOLECULAR FORCES

The subject of intermolecular forces is difficult and complex. In order to understand phenomena in solutions, however, it is necessary to recognize the following different types of intermolecular forces:

1. Forces between nonpolar molecules (London dispersion forces)
2. Repulsive forces
3. Coulombic forces (between ions)
4. Forces between permanent dipoles
5. Forces between ions and permanent dipoles
6. Forces between ions and induced dipoles
7. Forces between permanent and induced dipoles
8. Charge-transfer forces
9. Hydrogen bonding
10. Metallic bonding
11. Coordination bonding

In this chapter we will discuss briefly only those types of forces which are necessary for understanding the subsequent chapters.

1.3.1 London Dispersion Forces

The energy due to the London dispersion forces [3] is given by

$$u_{12} = -\frac{3\alpha_1\alpha_2}{2r^6}\frac{I_1 I_2}{I_1 + I_2} \tag{1.7a}$$

where α is the polarizability, r is the distance between the centers of the two molecules, and I is the ionization energy. If the two molecules are of the same species,

$$u_{11} = -\frac{3\alpha_1{}^2}{4r^6} I_1, \qquad u_{22} = -\frac{3\alpha_2{}^2}{4r^6} I_2 \tag{1.7b}$$

Accordingly,

$$|u_{12}| = \frac{2(I_1 I_2)^{1/2}}{I_1 + I_2}|u_{11}|^{1/2}|u_{22}|^{1/2} \tag{1.8}$$

In many cases, the ionization energies of the different molecules are not much different, and hence $2(I_1 I_2)^{1/2}/(I_1 + I_2) \approx 1$. Thus, the interaction energy between two different molecules is close to the geometrical mean of those of the respective molecules. Since geometric means are less than arithmetic means, larger differences in intermolecular potentials of the components of a mixture lead to a weakening of cohesion, i.e., less stabilization, endothermic mixing, and a positive heat of mixing.

The important features of this potential for the study of solution are that (1) it is temperature independent; (2) it is nonspecific; (3) it is short range (r^{-6}); (4) its strength depends on the number and "looseness" of the electrons, e.g., $I_2 > Br_2 > Cl_2 > F_2$; (5) it is close to the geometrical means of the potentials of the individual molecules.

1.3.2 *Repulsive Forces and Potential Energy Functions*

At small intermolecular distances, a repulsive potential abruptly sets in. This potential has been represented by an exponential function of the distance of separation r, but it is more frequently represented by a high inverse power, r^{-n}. The repulsion rises so quickly with diminishing separation that using n = 12 does not result in a serious discrepancy. In fact, for many properties good results are obtained by assuming the particles to be hard spheres, i.e., $n = \infty$.

Using n = 12, Lennard-Jones derived the potential energy function

$$u = 4\varepsilon\left[\left(\frac{\sigma}{r}\right)^{12} - \left(\frac{\sigma}{r}\right)^{6}\right] \tag{1.9}$$

where $-\varepsilon$ is the minimum potential energy and σ is the distance when u = 0. The intermolecular distance d_0 at which u is a minimum is related to σ by $\sigma = (2^{-1/6})d_0 = 0.8909d_0$.

1.3.3 *Dipole-Dipole Interactions*

The interaction between two permanent dipoles depends on their relative orientation. At very low temperatures, molecules tend to take up the most favorable orientation, but at moderate temperature, thermal agitation interferes with this orientation, and dipole-dipole attraction would disappear at sufficiently high temperatures, where the molecules orient in a completely random manner. Applying Boltzmann statistics, Keesom [4] derived the potential resulting from dipole-dipole interactions:

$$u_{12} = -\frac{2}{3}\frac{\mu_1^2\mu_2^2}{r^6 kT} \tag{1.10}$$

where μ is the dipole moment, r is the distance between the two dipoles, and k and T have their usual meanings. The important features of this potential are that (1) it is inversely proportional to the temperature; (2) it is short range, r^{-6}; (3) it is proportional to the square of the dipole moments of the respective molecules.

Dipole-dipole interactions or ion-dipole interactions lead to a stabilization and increase as the temperature decreases.

1.4 FREE VOLUME OF LIQUIDS

The term *free volume* has been loosely used for three distinguishable concepts. One is the excess of the volume occupied per mole of substance over the sum of the volumes of the individual molecules. A second is the excess of the molar volume over the volume at 0 K. The third is the volume calculated from the entropy of vaporization.

The simplest and most useful definition of free volume is in terms of the entropy of vaporization. From statistical thermodynamics [5]:

$$\frac{RT}{V} = P = \frac{RT}{V_f} \exp\left(\frac{-\Delta H^V}{RT}\right) = \frac{RT}{V_f} \exp\left(\frac{-\Delta S^V}{R}\right) \tag{1.11}$$

or

$$\Delta S^V = R \ln\left(\frac{V}{V_f}\right) \tag{1.12}$$

where V_f is the molal free volume, P is the equilibrium vapor pressure, and ΔH^V and ΔS^V are the heat and entropy of vaporization. As the entropy of vaporization is accurately determined, the free volume is readily calculated by means of Eq. (1.12). The same relation also can be derived thermodynamically. The total differential of the entropy is

$$dS = \left(\frac{\partial S}{\partial T}\right)_V dT + \left(\frac{\partial S}{\partial V}\right)_T dV \tag{1.13}$$

At constant temperature

$$dS = \left(\frac{\partial S}{\partial V}\right)_T dV \tag{1.14}$$

A real fluid is described by the van der Waals equation

$$\left(P + \frac{a}{V^2}\right)(V - b) = RT$$

Applying this relation to Eq. (1.14),

$$\left(\frac{\partial S}{\partial V}\right)_T = \left(\frac{\partial P}{\partial T}\right)_V = \frac{R}{V - b} \tag{1.15}$$

$$dS = \frac{R}{V - b}\, dV$$

Thus, the entropy of expansion of a van der Waals fluid from V_1 to V_2 would be

$$\Delta S = R \ln \frac{V_2 - b}{V_1 - b} \tag{1.16}$$

where $V - b$ is the molal free volume of the van der Waals fluid. If we define the free volume of the liquid, V_f, so as to satisfy Eq. (1.16), we obtain the same relation as in Eq. (1.12) for the process of vaporization.

Eyring and Hirschfelder [6] and subsequently Lennard-Jones and Devonshire [7] assumed that the center of each molecule moves in a spherical cage whose radius r is the difference between the average distance a between centers of nearest-neighbor molecules, and the diameter d of a molecule regarded as a rigid sphere (see Fig. 1.3).

The molal free volume is expressed by the equation

$$V_f = \frac{4}{3}\pi\gamma(a - d)^3 N_0 \tag{1.17}$$

where γ is a constant determined by the number of nearest neighbors, Z, i.e.,

$$\begin{aligned} Z &= 12, & \gamma &= \sqrt{2} \\ Z &= 8, & \gamma &= \sqrt{3/4} \\ Z &= 6, & \gamma &= 1 \\ Z &= 4, & \gamma &= 3\sqrt{3/8} \end{aligned} \tag{1.18}$$

For the cubic cage of a cubic quasilattice [8]

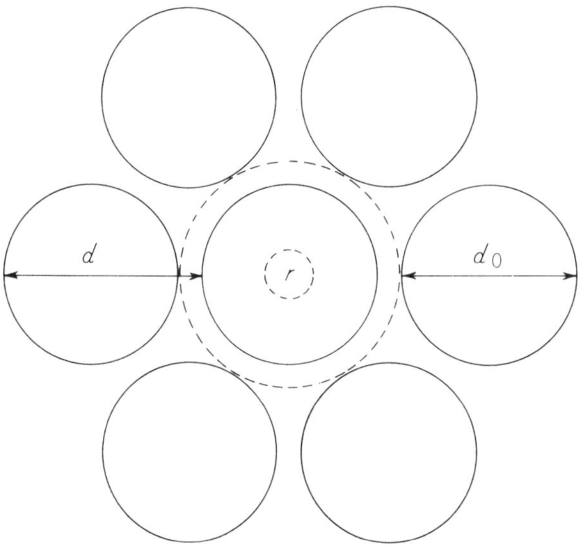

Fig. 1.3 Spherical cage model for free volume.

$$V_f = 8(a - d)^3 N_0 \tag{1.19}$$

The very small volumes calculated from the entropy of vaporization by means of Eq. (1.12) agree with the volumes swept out by the centers of molecules oscillating in the cages. Although the nearest neighbors of a molecule are not actually arranged in a geometrical lattice in the liquid state, and a molecule involved with its neighbors is in irregular, noiselike oscillations, this is an important correlation. We can calculate the entropy of mixing for liquids which mix in a completely random way, just as we can calculate the entropy of mixing of gases, provided we evaluate the free volumes of the respective components in solution.

1.5 MEASURES OF COMPOSITION

The composition of a solution can be expressed in a variety of ways.

1. *Mole fraction:* The mole fraction is the ratio of the number of moles of each component to the total number of moles:

$$x_1 = \frac{n_1}{n_1 + n_2 + \cdots} \tag{1.20}$$

2. *Mole ratio:* The mole ratio is the ratio of the number of moles of one component of a binary solution to the number of moles of the second component, i.e., n_1/n_2.

3. *Molality and molarity:* In the case of aqueous solutions it has become the almost universal custom to express numerical data either as the molality (m, moles per 1000 g of solvent) or as the molarity (M, moles per liter of solution). The molality of a solution is independent of the temperature, while the molarity is not.

4. *Volume fraction:* The volume fraction is the ratio of the volume of each component to the total volume:

$$\phi_1 = \frac{n_1 v_1}{n_1 v_1 + n_2 v_2 + \cdots} \tag{1.21}$$

It should be noted that the volume of a solution is not strictly equal to the sum of the volumes of its components, but is the fractional sum of their partial molal volumes. Volume fractions are extensively used in dealing with polymer solutions. Almost all theories and data disregard the volume change on mixing and the volume ratio change with temperature, because these are usually very small and not worth considering in practice.

5. *Weight fraction:* The weight fraction is the ratio of the weight of each component to the total weight.

There are, in addition, other ways of expressing composition, such as normality, dilution, and weight percent.

For the purposes of physical chemistry, mole fractions and volume fractions are most frequently used, while for analytical and many practical applications the other units described above may be employed.

1.6 PARTIAL MOLAL QUANTITIES [9]

It is assumed that the reader has an elementary acquaintance with thermodynamics and that it is necessary only to discuss the relations in multicomponent systems.

We may now consider characteristics common to all partial molal quantities, applications of which are important in analyzing the phenomena of solutions and solubility.

1.6.1 *General Partial Molal Equations*

Let Y be any extensive property of a given solution, such as entropy, volume, or heat capacity, which is a function of temperature, pressure, and the amounts of constituents. At constant temperature and pressure, Y depends only on n_1, n_2, ...

$$dY = \left(\frac{\partial Y}{\partial n_1}\right)_{P,T,n_2,n_3,\ldots} dn_1 + \left(\frac{\partial Y}{\partial n_2}\right)_{P,T,n_2,n_3,\ldots} dn_2 + \cdots \tag{1.22}$$

The partial molal quantities are defined by the equations

$$\bar{Y}_1 = \left(\frac{\partial Y}{\partial n_1}\right)_{P,T,n_2,n_3,\ldots}, \qquad \bar{Y}_2 = \left(\frac{\partial Y}{\partial n_2}\right)_{P,T,n_2,n_3,\ldots} \tag{1.23}$$

Equation (1.22) can then be written

$$dY = \bar{Y}_1\, dn_1 + \bar{Y}_2\, dn_2 + \cdots \tag{1.24}$$

Namely, the partial molal quantities are defined as the rate of change in the content of a particular quantity of the system while a particular component is being added at constant temperature and pressure. Partial molal quantities are designated by bars above the symbols for molal quantities, with the particular component designated by a subscript, e.g., $\bar{S}_1$, $\bar{S}_2$, ..., $\bar{G}_1$, $\bar{G}_2$, ..., $\Delta\bar{S}_1$, $\Delta\bar{S}_2$, ..., .

It is evident that partial molal quantities such as $\bar{Y}_1$ are not extensive, but intensive properties of the solution. They depend, therefore, not on the total amount of each constituent, but only on the composition. If, therefore, we add several constituents simultaneously to a solution at constant temperature and pressure, keeping their ratios constant, the partial molal quantities will be unchanged. We may therefore integrate Eq. (1.24), keeping n_1, n_2, ..., in constant proportions, and find

$$Y = \bar{Y}_1 n_1 + \bar{Y}_2 n_2 + \cdots \tag{1.25}$$

In deriving Eq. (1.25) we did not limit ourselves to any special values of n_1, n_2, ... Hence this equation, being entirely general, can be differentiated with respect to any change of composition. The differentiation of Eq. (1.25) gives

$$dY = \bar{Y}_1\, dn_1 + n_1\, d\bar{Y}_1 + \bar{Y}_2\, dn_2 + n_2\, d\bar{Y}_2 + \cdots \tag{1.26}$$

From Eqs. (1.24) and (1.26) we obtain

$$n_1\, d\bar{Y}_1 + n_2\, d\bar{Y}_2 + \cdots = 0 \tag{1.27}$$

For 1 mol of solution,

$$x_1\, d\bar{Y}_1 + x_2\, d\bar{Y}_2 + \cdots = 0 \tag{1.28}$$

$$Y = x_1 \bar{Y}_1 + x_2 \bar{Y}_2 + \cdots \tag{1.29}$$

Equations (1.27) and (1.28) show the relation between the change in any one $\bar{Y}_1$ and thc change in all the others for any infinitesimal alteration in composition at constant temperature and pressure. We may regard the number of moles of one constituent, say n_2, as the main variable. Equation (1.27) then takes the form

$$n_1 \left(\frac{\partial \bar{Y}_1}{\partial n_2}\right)_{P,T} + n_2 \left(\frac{\partial \bar{Y}_2}{\partial n_2}\right)_{P,T} + \cdots = 0 \tag{1.30}$$

Similarly, Eq. (1.28) becomes

$$x_1 \left(\frac{\partial \bar{Y}_1}{\partial x_1}\right)_{P,T} + x_x \left(\frac{\partial \bar{Y}_2}{\partial x_1}\right)_{P,T} + \cdots = 0 \tag{1.31}$$

It will be noted that while n_1 can vary without any change in n_2, n_3, ..., we cannot change x_1 without a change in x_2, x_3, ...

Thus, in a binary solution

$$dx_1 = -dx_2$$

If we apply Eq. (1.31) to a binary solution,

$$x_1\left(\frac{\partial \bar{Y}_1}{\partial x_1}\right)_{P,T} + x_2\left(\frac{\partial \bar{Y}_2}{\partial x_1}\right)_{P,T} = 0, \quad x_1\left(\frac{\partial \bar{Y}_1}{\partial x_1}\right) - x_2\left(\frac{\partial \bar{Y}_2}{\partial x_2}\right) = 0 \tag{1.32}$$

or

$$\frac{\partial \bar{Y}_1/\partial x_2}{\partial \bar{Y}_2/\partial x_2} = -\frac{x_2}{x_1} = -\frac{x_2}{1 - x_2} \tag{1.33}$$

If $\bar{Y}_1$ and $\bar{Y}_2$ are plotted against x_2, the two differential coefficients are the slopes of the two curves, and therefore the slope of one curve is entirely determined by the slope of the other, and by the composition. Thus, at 50 mol% where $x_1 = x_2$, the slopes of the two curves must be equal and opposite in sign. If, at any composition, one of the curves has a maximum, the other curve will have a minimum at the same composition, as shown in Figs. 1.4 and 1.5.

Thermodynamic relations which hold for the entire solution or for one-component systems, hold equally well for partial molal quantities. For example,

$$\left(\frac{\partial G_2}{\partial P}\right)_{T,n_1} = \frac{\partial}{\partial P}\left(\frac{\partial G}{\partial n_2}\right)_{P,T,n_1} = \frac{\partial}{\partial n_2}\left(\frac{\partial G}{\partial P}\right)_{P,T,n_1}$$

$$= \left(\frac{\partial V}{\partial n_2}\right)_{P,T,n_1} = \bar{V}_2 \tag{1.34}$$

Equations (1.24), (1.25), (1.27), and (1.34) demonstrate important characteristics common to all partial molal quantities.

1.6.2 Partial Molal Free Energy

The partial molal free energy of a component is often called the chemical potential, and it is designated by the symbol μ,

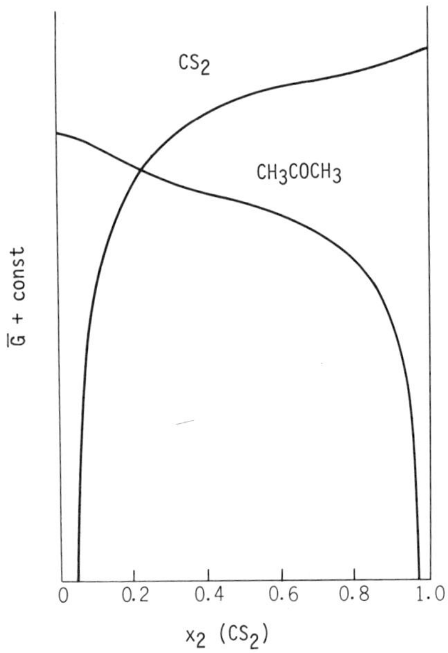

Fig. 1.4 Partial molal Gibbs free energies in solutions of carbon disulfide and acetone.

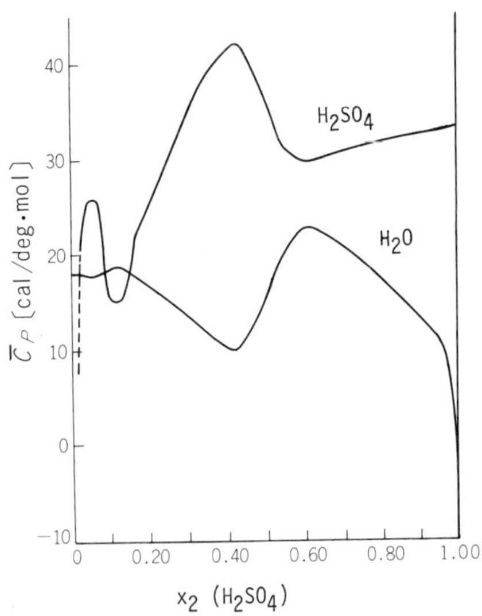

Fig. 1.5 Partial molal heat capacities in aqueous solutions of sulfuric acid.

$$\mu = \bar{G}_i = \left(\frac{\partial G}{\partial n_i}\right)_{P,T,n_j} \tag{1.35}$$

where all the n_j are constant, except n_i. The total differential of the free energy is

$$dG = \left(\frac{\partial G}{\partial T}\right)_{P,n_i} dT + \left(\frac{\partial G}{\partial P}\right)_{T,n_i} dP + \sum\left(\frac{\partial G}{\partial n_i}\right)_{P,T,n_j} dn_i$$

$$= S\ dT + V\ dP + \Sigma\mu_i\ dn_i \tag{1.36}$$

The solutions we are interested in usually are subject to constant temperature and pressure, so that the partial molal Gibbs free energy (= chemical potential) governs the transfer of the substances. At equilibrium $\mu_i^1 = \mu_i^2$ holds. Otherwise components transfer in the direction of lower chemical potential. Occasionally, we deal with a system under constant volume and temperature. Similar theoretical considerations may be applied for these conditions. Here, it is convenient to consider the Helmholtz free energy, rather than the Gibbs free energy.

$$dA = dG - P\ dV - V\ dP$$

Substituting Eq. (1.36)

$$dA = S\ dT - P\ dV + \sum_i \mu_i\ dn_i \tag{1.37}$$

Since Eq. (1.37) is the total differential of A,

$$\mu_i = \left(\frac{\partial A}{\partial n_i}\right)_{T,V,n_j} = \left(\frac{\partial G}{\partial n_i}\right)_{T,P,n_j} \tag{1.38}$$

The condition for equilibrium at constant temperature and volume is dA = 0. Thus, we know that the same chemical potentials govern the transfer of substances for constant volume and for constant pressure conditions.

1.6.3 *The Fugacity of a Dissolved Substance*

Just as in the case of pure substances, the fugacity (or escaping tendency) of a solution constituent is defined by the equation

$$\left(\frac{\partial \ln f_1}{\partial \mu_1}\right)_T = \frac{1}{RT} \tag{1.39}$$

or

$$\mu_1 = RT \ln f_1 = \mu_1^*(T) \tag{1.40}$$

where

$$\mu_1^*(T) = \lim_{P_1 \to 0} (\mu_1 - RT \ln p_1)$$

Hence $f_1 = p_1$ at low pressure.

The total differential of the chemical potential of component 1 in the gas phase is

$$d\mu_1{}^g = RT\ d \ln f_1 + R \ln f_1\ dT - S_1^*\ dT \tag{1.41}$$

The total differential of the chemical potential of component 1 in solution is derived from Eq. (1.36)

$$d\mu_1{}^\ell = -S_1{}^\ell\ dT + V_1{}^\ell\ dP + \mu_1\ dx_1 \tag{1.42}$$

Substituting the relation

$$\mu_1{}^\ell = \bar{H}_1{}^\ell - T\bar{S}_1{}^\ell, \qquad \mu_1^* = H_1^* - TS_1^*$$

into Eq. (1.40), we obtain

$$R \ln f_1 = \frac{\bar{H}_1{}^\ell - H_1^*}{T} - \bar{S}_1{}^\ell - S_1^* \tag{1.43}$$

Equating Eqs. (1.41) and (1.42), and introducing the relation (1.43)

$$d \ln f_1 = \frac{H_1^* - \bar{H}_1{}^\ell}{RT^2}\ dT + \frac{\bar{V}_1{}^\ell}{RT}\ dP + \frac{\mu_1}{RT}\ dx_1 \tag{1.44}$$

Differentiating Eq. (1.44) with respect to P and T, we obtain, respectively

$$\left(\frac{\partial \ln f_1}{\partial P}\right)_{T,x} = \frac{\bar{V}_1^{\ell}}{RT} \tag{1.45}$$

and

$$\left(\frac{\partial \ln f_1}{\partial T}\right)_{P,x} = \frac{H_1^* - \bar{H}_1^{\ell}}{RT^2} \tag{1.46}$$

Thus, we find that the effect of pressure and temperature on the escaping tendencies of the components of a solution are identical in form to those found for pure substances.

1.6.4 The Gibbs-Duhem Equation

Introducing $\bar{G}_1$ into Eq. (1.31), we obtain

$$x_1\left(\frac{\partial \bar{G}_1}{\partial x_1}\right)_{P,T} + x_2\left(\frac{\partial \bar{G}_2}{\partial x_1}\right)_{P,T} + \cdots = 0 \tag{1.47}$$

In a binary solution

$$x_1\left(\frac{\partial \bar{G}_1}{\partial x_1}\right)_{P,T} + x_2\left(\frac{\partial \bar{G}_2}{\partial x_1}\right)_{P,T} = 0 \tag{1.48}$$

This important equation was first obtained by Gibbs in 1875. Equation (1.48) and its equivalent forms are called the Gibbs-Duhem equation. The behavior of the partial molal free energy of each constituent in a mixture of carbon disulfide and acetone is shown in Fig. 1.4

The simplest, and on the whole the most useful, method in the determination of partial molal Gibbs free energy consists in determining the vapor pressure ($\approx$ fugacity) of the constituents of a solution.

If we use fugacity in place of free energy, Eq. (1.47) becomes

$$x_1\left(\frac{\partial \ln f_1}{\partial x_1}\right)_{P,T} + x_2\left(\frac{\partial \ln f_2}{\partial x_1}\right)_{P,T} + \cdots = 0 \tag{1.49}$$

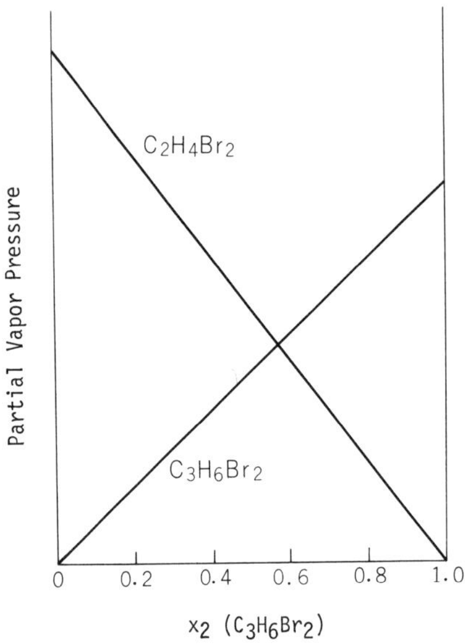

Fig. 1.6 Partial vapor pressures for solutions of ethylene dibromide and propylene dibromide.

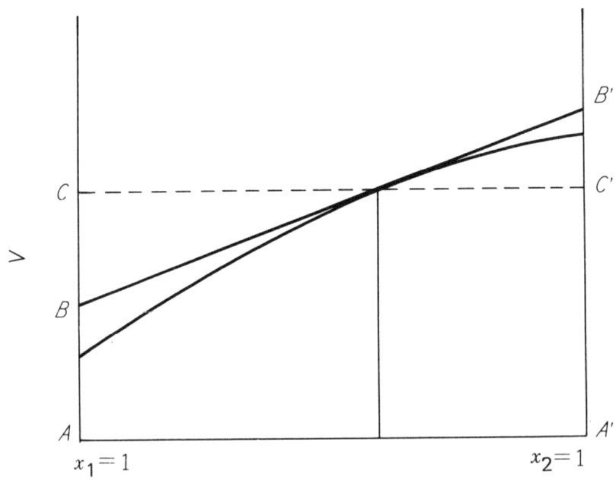

Fig. 1.7 The relations among molal volumes of solutions and their partial molal volumes.

In a binary solution, where $dx_1 = -dx_2$

$$x_1\left(\frac{\partial \ln f_1}{\partial x_1}\right)_{P,T} = x_2\left(\frac{\partial \ln f_2}{\partial x_2}\right)_{P,T} \tag{1.50}$$

$$\left(\frac{\partial \ln f_1}{\partial \ln x_1}\right)_{P,T} = \left(\frac{\partial \ln f_2}{\partial \ln x_2}\right)_{P,T}$$

When the vapors are nearly perfect gases, we may substitute partial pressures for fugacities and obtain the approximate relation

$$x_1\left(\frac{\partial \ln P_1}{\partial x_1}\right)_{P,T} = x_2\left(\frac{\partial \ln P_2}{\partial x_2}\right)_{P,T} \tag{1.51}$$

This approximate equation was found by Duhem and Margules and it is in this form that the relations between fugacity and composition were verified.

Equation (1.51) may be put in the alternative form

$$\frac{(\partial P_1/\partial x_1)_T}{(\partial P_2/\partial x_2)_T} = \frac{P_1/x_1}{P_2/x_2} \tag{1.52}$$

If one of the curves is a straight line over the whole composition range (e.g., $\partial P_1/\partial x_1 = P_1/x_1$), the other is also a straight line over the same range, as shown in Fig. 1.6.

1.6.5 One Graphical Method of Determining Partial Molal Quantities

There is a simple graphical method for determining partial molal quantities from a plot of the total quantity per mole of mixture. The variation of V with composition is represented in Fig. 1.7. The slope of the tangent to the curve at x_1 is

$$\frac{dV}{dx_2} = \left(\frac{\partial V}{\partial x_2}\right)_{x_1} - \left(\frac{\partial V}{\partial x_1}\right)_{x_2} = \bar{V}_2 - \bar{V}_1 \tag{1.53}$$

From Eq. (1.29)

$$V = x_1\bar{V}_1 + x_2\bar{V}_2 = AC \tag{1.54}$$

It is evident that

$$AB = AC - BC = x_1\bar{V}_1 + x_2\bar{V}_2 - (\bar{V}_2 - \bar{V}_1)x_2 = \bar{V}_1$$

$$A'B' = A'C' + B'C' = x_1\bar{V}_1 + x_2\bar{V}_2 + (\bar{V}_2 - \bar{V}_1)x_1 = \bar{V}_2$$

This result is a perfectly general one for any extensive property of a binary mixture or of any of the "excess" functions, i.e., increments over ideal mixing. This graphic relation clearly illustrates the very important concept that a partial molal quantity is really a property of the solution *as a whole*, not a property of the component in question.

For example, the partial molal volume of a solute must not be looked on as the volume occupied by a mole of solute. The partial molal volume is the total volume increase when solute molecules are added; a major part of this increase is surely due to a change in the packing of the adjacent solvent molecules.

In the discussion of solubility, the partial molal enthalpy and entropy of the solute determine the solubility, i.e., (1) the partial molal quantities are basic to the study of solubility, and (2) the change in the thermodynamic properties of the solvent molecules directly affects the solubility of the solute molecule. In the case of dissolution in water, the enthalpy and entropy of solution of the solute markedly decreases, but the free energy does not change greatly.

REFERENCES

1. D. Cook, Trans. Faraday Soc., *49*, 716 (1953).
2. J. H. Hildebrand, J. Am. Chem. Soc., *37*, 970 (1915); *40*, 45 (1918).
3. R. Eisenschitz and F. London, Z. Phys., *60*, 49 (1930); F. London, Z. Phys., *63*, 245 (1930); Z. Phys. Chem. B, *11*, 222 (1930); Trans. Faraday Soc., *33*, 8 (1937).
4. W. H. Keesom, Phys. Z., *22*, 126 (1921); *22*, 643 (1921); *23*, 225 (1922).

5. R. H. Fowler and E. A. Guggenheim, *Statistical Thermodynamics*, Cambridge University Press, 1939, p. 333.
6. H. Eyring and J. Hirschfelder, J. Phys. Chem., *41*, 249 (1937).
7. J. E. Lennard-Jones and A. F. Devonshire, Proc. Roy. Soc. A, *163*, 59 (1937).
8. J. O. Hirschfelder, C. F. Curtiss, and R. B. Bird, *Molecular Theory of Gases and Liquids*, Wiley, New York, 1954, p. 280.
9. G. N. Lewis and M. Randall, *Thermodynamics*, 2d ed. revised by K. S. Pitzer and L. Brewer, McGraw-Hill, New York, 1961.

EXERCISES

1.1 By further partial differentiation of

$$\left(\frac{\partial E}{\partial V}\right)_T = T\left(\frac{\partial P}{\partial T}\right)_V - P$$

obtain the following equation, and interpret its meaning.

$$\left(\frac{\partial C_V}{\partial V}\right)_T = T\left(\frac{\partial^2 P}{\partial T^2}\right)_V$$

Answer:

$$\left(\frac{\partial C_V}{\partial V}\right)_T = \left(\frac{\partial^2 E}{\partial V \partial T}\right) = \left(\frac{\partial^2 E}{\partial T \partial V}\right) = \left(\frac{\partial P}{\partial T}\right)_V + T\left(\frac{\partial^2 P}{\partial T^2}\right)_V - \left(\frac{\partial P}{\partial T}\right)_V = T\left(\frac{\partial^2 P}{\partial T^2}\right)$$

This means that what would be a very difficult measurement experimentally, can be reduced to the measurement of a much more readily measurable quantity.

1.2 A constant-boiling solution is one in which evaporation causes no change in the composition of the liquid. In other words, the composition of the liquid and gaseous phases are identical. If the vapors may be assumed to be ideal gases, then the ratio of the two partial pressures is equal to the ratio of the mole fraction in the liquid. Hence, show that

$$\frac{dp_1}{dx_1} + \frac{dp_2}{dx_1} = 0$$

and that the total vapor pressure $P = p_1 + p_2$ is a maximum or minimum. This is an important principle in the theory of distillation.

Answer: The Gibbs-Duhem equation (1.50) is

$$x_1\left(\frac{\partial \ln f_1}{\partial x_1}\right)_T + x_2\left(\frac{\partial \ln f_2}{\partial x_1}\right)_T = 0$$

When the vapors behave ideally

$$\frac{x_1}{p_1}\left(\frac{\partial p_1}{\partial x_1}\right)_T + \frac{x_2}{p_2}\left(\frac{\partial p_2}{\partial x_1}\right)_T = 0$$

In a constant-boiling (azeotropic) mixture $x_1/x_2 = p_1/p_2$, hence

$$\left(\frac{\partial p_1}{\partial x_1}\right)_T + \left(\frac{\partial p_2}{\partial x_2}\right)_T = 0 = \left(\frac{\partial P}{\partial x_1}\right)_T$$

Therefore, P is either a maximum or minimum.

1.3 Derive the following relations

$$\sum n_i\left(\frac{\partial \ln f_1}{\partial P}\right)_{T,n_j} = \frac{V}{RT}$$

$$\sum n_i\left(\frac{\partial \ln f_1}{\partial T}\right)_{P,n_j} = \frac{H^* - H}{RT^2}$$

where V is the volume and $H^* - H$ is the ideal heat of vaporization of the solution.

Answer: From Eqs. (1.45) and (1.25)

$$\left(\frac{\partial \ln f_1}{\partial P}\right)_{T,n} = \frac{\bar{V}_1}{RT}$$

$$\frac{V}{RT} = \frac{1}{RT}\sum n_i\bar{V}_i = n_1\left(\frac{\partial \ln f_1}{\partial P}\right) + n_2\left(\frac{\partial \ln f_2}{\partial P}\right) + \cdots$$

From Eqs. (1.46) and (1.25)

$$\left(\frac{\partial \ln f_1}{\partial T}\right)_{P,n} = \frac{H_1^* - H_1}{RT^2}$$

$$\sum n_i \left(\frac{\partial \ln f_1}{\partial T} \right) = \sum n_i \left(\frac{H_1^* - \bar{H}_1}{RT^2} \right) = \frac{H^* - H}{RT^2}$$

PROBLEMS

1.1 Express the following 35 thermodynamic relations in terms of C_p, C_v, P, V, T, S, $(\partial p/\partial T)_v$, $(\partial V/\partial T)_p$, $(\partial p/\partial V)_T$.

x =	G	H	S	E	F
$(\partial x/\partial T)_V$ =					
$(\partial x/\partial T)_P$ =					
$(\partial x/\partial V)_T$ =					
$(\partial x/\partial P)_T$ =					
$(\partial x/\partial P)_V$ =					
$(\partial x/\partial V)_P$ =					
$(\partial x/\partial T)_S$ =					

1.2 In a mixture of water and ethanol in which the mole fraction of water is 0.4, the partial molal volume of the alcohol is 57.5 cm^3, and the density of the mixture is 0.8494, calculate the partial molal volume of the water.

1.3 Show that maximum in the curve of a partial molal quantity versus composition for one component of a binary mixture must come at the same composition as a minimum for the other.

chapter two

IDEAL SOLUTIONS AND ATHERMAL SOLUTIONS

2.1 IDEAL SOLUTIONS

In order to understand any property of matter, it is convenient to start from an ideal property. Just as the ideal gas laws suffice to describe the behavior of actual gases as a first approximation, the laws of ideal solutions suffice to describe many actual solutions within certain limitations.

When two liquids with similar physical and chemical properties are mixed, it is observed that the volume and energy change of mixing are practically zero. The partial vapor pressures of respective components in such a solution are proportional to their mole fraction in solution. Such experimental observations may be explained by various approaches.

For example, let us consider the kinetic theory. When the molecules of two different constituents are nearly identical it may be expected that the force acting between unlike molecules would be almost equal to the force acting between similar molecules, and that the number of molecules of either constituent escaping into the vapor phase would be proportional to the relative number of those molecules in the liquid and at the surface.

It would be expected that isomers of organic compounds, e.g., hydrocarbons with different branching, may show such similar properties that the partial pressures and mole fractions would be proportional to each other. Thus, there should be no measurable departure from a linear relation. In the case of isotopes, which are nearly identical, the properties of the solution may be readily predicted from those of the constituents, to a close approximation. On the other hand, the separation of the respective components by the use of different properties becomes extremely difficult. A perfect, or ideal, solution may be considered as the extreme extrapolation of this situation.

Ideal behavior may be (1) defined thermodynamically in terms of empirical equations, or (2) derived from ideal models on the basis of molecular theory.

An ideal solution is defined as one in which the mixing of two liquids which consist of equal-sized molecules is not accompanied by heat and volume changes. In such a system, the enthalpy of each component remains unchanged, since no heat of mixing is involved. Thus, the decrease in Gibbs energy is entirely a result of the increase in entropy.

$$\Delta H^M = 0, \qquad \Delta\bar{H}_2 = 0 \tag{2.1}$$

$$\Delta G^M = -T\Delta S^M, \qquad \Delta\bar{G}_2 = -T\Delta\bar{S}_2 \tag{2.2}$$

The relation applies to the properties of the total solution as well as to the partial molal quantities, since the relation holds over the whole composition range. When no change of volume is involved and no heat of solution is required, the two types of molecules will mix in complete randomness and the molal volume and the free volume of the molecules should be the same before and after mixing. Accordingly, when n_1 moles of the first component is mixed with n_2 moles of the second component, the free volume for the two components increases from n_1V_f and n_2V_f, respectively, to $(n_1 + n_2)V_f$ for each component. Thus, the increase in the entropy of the system may be expressed as follows, in analogy to the case of mixing two ideal gases at the same pressure, i.e., same free volume, at constant temperature:

$$\Delta S^M = n_1 R \ln \frac{(n_1 + n_2)V_f}{n_1 V_f} + n_2 R \ln \frac{(n_1 + n_2)V_f}{n_2 V_2} \tag{2.3}$$

which may be simplified to

$$\Delta S^M = -R\left(n_1 \ln \frac{n_1}{n_1 + n_2} + n_2 \ln \frac{n_2}{n_1 + n_2}\right) \tag{2.4}$$

From statistical thermodynamics, the number of possible combinations increases by $(N_1 + N_2)!/N_1!N_2!$ when N_1 molecules of A and N_2 molecules of B are mixed in a perfectly random manner in the volume occupied by the $N_1 + N_2$ molecules. The increase in entropy can then be calculated from the Boltzmann equation [1]:

$$\Delta S^M = k \ln\left[\frac{(N_1 + N_2)!}{N_1!N_2!}\right]$$

Since $N_1 = n_1 N_0$, $N_2 = n_2 N_0$, for large N_1 and N_2, this becomes

$$\Delta S^M = -kN_0\left(n_1 \ln \frac{n_1}{n_1 + n_2} + n_2 \ln \frac{n_2}{n_1 + n_2}\right) \tag{2.5}$$

where N_0 is Avogadro's number. Equation (2.5) is identical to Eq. (2.4). Hence, the same conclusion can be drawn from the statistical argument as from the thermodynamic one.

Since the entropy of mixing is expressed by Eq. (2.5), the partial molal entropy of mixing of the second component is

$$\Delta \bar{S}_2 = \left(\frac{\partial \Delta S^M}{\partial n_2}\right)_{n_1} = -R \ln \frac{n_2}{n_1 + n_2} = -R \ln x_2 \tag{2.6}$$

The partial molal entropy of mixing of each component and the total entropy of mixing of 1 mol of ideal solutions are shown in Fig. 2.1.

The Gibbs free energy of mixing of the solutions is given from Eqs. (2.2) and (2.4):

$$\Delta G^M = RT(n_1 \ln x_1 + n_2 \ln x_2) \tag{2.7}$$

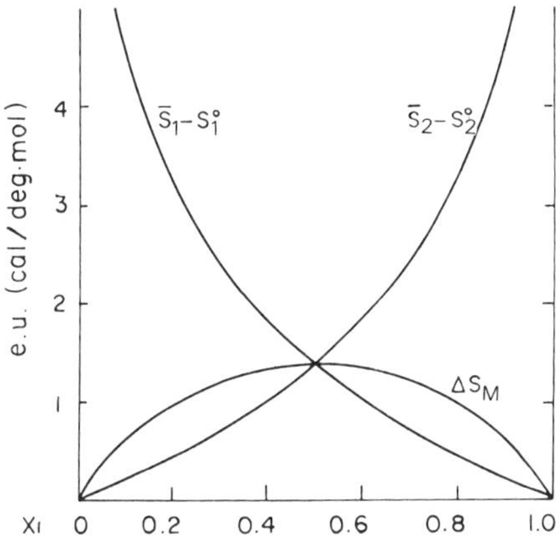

Fig. 2.1 Partial and total entropies of mixing in ideal solutions.

The change in the partial molal Gibbs free energy of the second component due to the mixing is

$$\Delta\bar{G}_2 = RT \ln x_2 \tag{2.8}$$

If vapor pressures are substituted as a close approximation to fugacities, the molal Gibbs free energy of the second component is

$$G_2^\circ = RT \ln P_2^\circ + B \tag{2.9}$$

where B is the constant which, for a given temperature and a given substance, makes the fugacity equal to the pressure in the ideal gas state. Similarly, the partial molar free energy in an ideal solution of mole fraction x_2 is

$$G_2^\circ + \Delta\bar{G}_2 = \bar{G}_2 = RT \ln P_2^\circ x_2 + B \tag{2.10}$$

Thus, we see that the partial pressure of the component is proportional to the mole fraction.

The partial and total Gibbs free energies of mixing of 1 mol of ideal solutions are shown in Fig. 2.2.

Although ideal solutions do not exist, the above relation almost holds for solutions consisting of isotopes, isomers, and homologues, as well as the systems n-C_7H_{16}/$CH_3C_6H_{11}$, C_6H_6/$C_2H_4Cl_2$, CH_3I/$CHCl_3$, $SnCl_4$/CCl_4, and I_2/SnI_4. These examples include cases

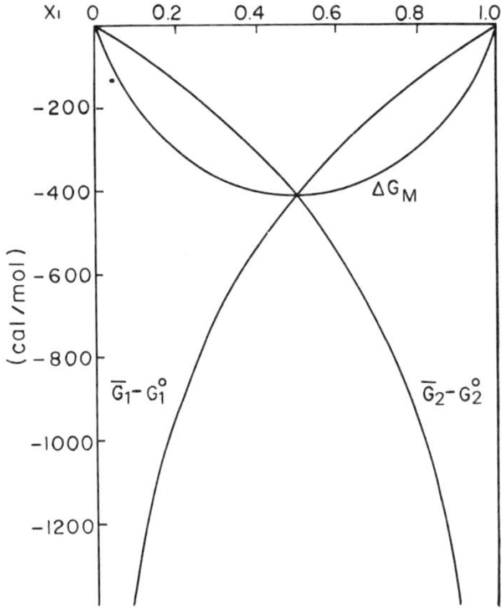

Fig. 2.2 Partial and total Gibbs free energies of mixing in ideal solutions.

where the molar volumes of the two components are somewhat different.

An ideal solution also may be defined as one in which the fugacity of each component is proportional to its mole fraction, and in which the proportional relationship holds over all ranges of pressure and temperature. In such a case, for the first component

$$f_1 = f_1^\circ x_1 \tag{2.11}$$

where f_1° is the fugacity of the first component in the pure liquid state and is a constant at a given temperature and pressure. Accordingly,

$$\bar{G}_1 - G_1^\circ = RT \ln x_1 \tag{2.12}$$

From the previous definition of an ideal solution, Eq. (2.12) should hold at any temperature and pressure. Differentiating Eq. (2.12) with respect to pressure, we obtain

$$\left(\frac{\partial \bar{G}_1}{\partial P}\right)_T - \left(\frac{\partial G_1^\circ}{\partial P}\right)_T = \bar{V}_1 - V_1^\circ = 0 \tag{2.13}$$

Thus, the partial molar volume of the first component is equal to the molar volume of the pure component V_1° at all compositions. Since the corresponding equations apply to all components, there can be no change in volume on mixing for an ideal solution.

Similarly we may differentiate Eq. (2.12) with respect to temperature,

$$\left(\frac{\partial G_1}{\partial T}\right)_P - \left(\frac{\partial G_1^\circ}{\partial T}\right)_P = -\bar{S}_1 + S_1^\circ = R \ln x_1 \tag{2.14}$$

Combining Eqs. (2.12) and (2.14),

$$\bar{H}_1 - H_1^\circ = \bar{G}_1 - G_1^\circ + T(\bar{S}_1 - S_1^\circ) = 0 \tag{2.15}$$

Thus, the heat of mixing is seen to be equal to zero.

2.2 IDEAL SOLUTION IN EQUILIBRIUM WITH PURE SOLID PHASE

Let us consider the case of an ideal solution with a pure solid phase in equilibrium. A solution of benzene and naphthalene may be taken as an example. They form an almost ideal solution, but before the highest concentration of naphthalene is reached, solid naphthalene separates out. In an ideal solution the partial molal volume and the partial molal enthalpy of the dissolved naphthalene may be taken as equal to the molal volume and the molal enthalpy of supercooled liquid naphthalene. If f_2° is the fugacity of the supercooled liquid naphthalene, then, for any mole fraction of naphthalene, $f_2 = f_2^\circ x_2$. As the concentration of naphthalene in the solution increases, it crystallizes out when the partial pressure of the naphthalene equals the vapor pressure of solid naphthalene. Although the solution is ideal, the heat of mixing is not zero but equal to the heat of fusion of the solid.

If the heat of fusion of a solid (endothermic) cancels the exothermic heat of mixing, e.g., the heat of solvation, in a nonideal solution, the apparent heat of solution may become zero or small, but the solution could be quite nonideal.

2.3 ATHERMAL SOLUTIONS

A solution consisting of mixtures of molecules of different sizes, but having no heat of mixing, is called an athermal solution. In athermal solutions the sizes and the free volumes of the two constituent molecules may be different. The entropy of mixing is expressed in terms of the free volume in just the same way as for the mixing of ideal solutions.

$$\Delta S^M = -R\left[n_1 \ln \frac{n_1 V_{1f}}{n_1 \bar{V}_{1f} + n_2 \bar{V}_{2f}} + n_2 \ln \frac{n_2 V_{2f}}{n_1 \bar{V}_{1f} + n_2 \bar{V}_{2f}}\right] \tag{2.16}$$

where V_{1f} and V_{2f} are the molal free volumes and $\bar{V}_{1f}$ and $\bar{V}_{2f}$ are the partial molal free volumes of each constituent, respectively. Two limiting cases may be considered, depending on the relation between the free volumes of the constituents.

1. Suppose that the free volumes of two different molecules in the pure liquid and in solution are equal. Then Eq. (2.16) becomes identical with Eq. (2.4), which was obtained for ideal solutions. By partial differentiation of Eq. (2.16) with respect to each constituent, the partial molal entropy of mixing,

$$\Delta\bar{S}_2 = \left(\frac{\partial \Delta S^M}{\partial n_2}\right)_{n_1} = -R \ln x_2 \tag{2.17}$$

is obtained.

2. If the free volume of each constituent is proportional to the molal volume (as the other extreme assumption), we obtain [2]

$$\Delta\bar{S}_2 = -R\left[\ln \phi_2 + \phi_1\left(1 - \frac{V_2}{V_1}\right)\right] \tag{2.18}$$

where ϕ_1 and ϕ_2 are the volume fractions of each constituent, respectively. Equation (2.18) is identical with the Flory-Huggins equation for entropy of mixing in polymer solutions, described in Chap. 8. However, it is clear from the definition of the free volume of liquids that the free volume is not actually proportional to the molal volume, but rather is *nearly* equal when an athermal solution is formed. Thus, the first assumption seems closer to the facts in ordinary solutions. This concept will be discussed and verified in Chap. 6.

2.4 NONIDEAL SOLUTIONS

As the partial vapor pressure of each component is proportional to the mole fraction in an ideal solution, the relative activity of each component is equal to the mole fraction

$$a_2 = \frac{f_2}{f_2^\circ} = \frac{p_2}{p_2^\circ} = x_2 \tag{2.19}$$

Figure 2.3 indicates the relationship between the relative activity of each component and the mole fraction in ideal solutions.

The activity a of a component in a particular solution is the ratio of its fugacity to its fugacity in some arbitrarily chosen "standard state." The best choice for a standard state is the pure liquid, extrapolated below the melting point when dealing with the solubility of solids, and extrapolated to estimate the hypothetical

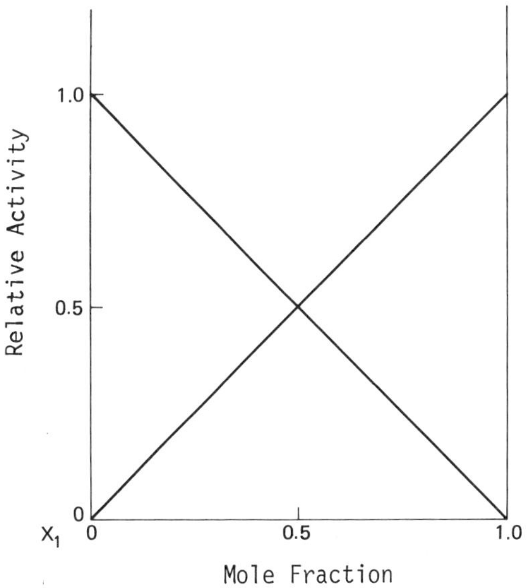

Fig. 2.3 Relative activities in ideal solutions.

fugacity of the pure liquid when dealing with the solubility of gases [3]. This convention yields equations of the same form for all components, and avoids a distinction, often quite arbitrary, between *solvent* and *solute*. In addition, this standard state makes the analysis of the behavior of solutions more readily comprehensible. In dealing with aqueous solutions of electrolytes, it is customary to adopt the infinitely dilute solution as the standard state, because of the difficulty of determining the relative activity of the solid electrolyte with respect to the pure liquid. The fact that activity so defined is different, depending on the types of electrolyte and additives to the solvent, presents serious difficulties. In the case of many organic electrolytes or bioelectrolytes, which are the substances not well explored and are more promising, a hydrated liquid state could be taken as the standard state. Whenever it is possible to evaluate the relative activity with respect to the pure liquid, the pure liquid should be adopted as the standard. In this book the standard state is invariably the pure liquid.

The relative activity of a component is not equal to its mole fraction in nonideal solutions. It becomes convenient to define the activity coefficients γ of a component as

$$\gamma_1 = \frac{a_1}{x_1}, \qquad \gamma_2 = \frac{a_2}{x_2} \tag{2.20}$$

The activity coefficient indicates the degree of deviation from ideal solution.

The partial molal Gibbs free energy of mixing is then expressed

$$\Delta\bar{G}_2 = \bar{G}_2 - \bar{G}_2^\circ = RT \ln a_2 = RT \ln x_2 + RT \ln \gamma_2 \tag{2.21}$$

The first term in Eq. (2.21) is the ideal free energy of mixing; thus, the second term corresponds to the excess due to the nonideal behavior of the system. The second term is called the *excess free energy*.

The excess partial molal free energy $\Delta\bar{G}_2{}^E$ can then be written

$$\Delta\bar{G}_2^{\ E} = RT \ln \gamma_2 \tag{2.22}$$

$$\Delta G^E = RT(x_1 \ln \gamma_1 + x_2 \ln \gamma_2) \tag{2.23}$$

Depending on whether $\gamma > 1$ or $\gamma < 1$, the deviations are called positive or negative, respectively. Other excess functions including entropy, enthalpy, and volume can be defined in a similar and consistent way. Since an ideal solution is formed with no heat and no volume of mixing, these excess functions are just the mixing functions themselves.

$$\bar{H}_1^{\ E} = H_1 - \bar{H}_1^{\ i} = \Delta\bar{H}_1 = -RT^2\left(\frac{\partial \ln \gamma_1}{\partial T}\right)_{p,x} = \left[\frac{\partial(\bar{G}_1^{\ E}/T)}{\partial(1/T)}\right]_{p,x} \tag{2.24}$$

$$\bar{S}_1^{\ E} = \bar{S}_1 - \bar{S}_1^{\ i} = -R\left(\frac{\partial \ln \gamma_1}{\partial \ln T} + \ln \gamma_1\right)_{p,x} = -\left(\frac{\partial\bar{G}_1^{\ E}}{\partial T}\right)_{p,x} \tag{2.25}$$

$$\bar{V}_1^{\ E} = \bar{V}_1 - \bar{V}_1^{\ i} = \Delta\bar{V}_1 = RT\left(\frac{\partial \ln \gamma_1}{\partial P}\right)_{T,x} = \left(\frac{\partial\bar{G}_1^{\ E}}{\partial P}\right)_{T,x} \tag{2.26}$$

where superscripts E and i mean excess and ideal, respectively.

REFERENCES

1. R. H. Fowler and E. A. Guggenheim, *Statistical Thermodynamics*, Cambridge University Press, England, 1939, pp. 353-354.
2. J. H. Hildebrand, J. Chem. Phys., *15*, 225 (1947).
3. J. H. Hildebrand, J. M. Prausnitz, and R. L. Scott, *Regular and Related Solutions*, Van Nostrand Reinhold, New York, 1970.

EXERCISES

2.1 Calculate the minimum work necessary to separate 1 mol of pure benzene from (a) a large volume of an equimolal solution of benzene and toluene; (b) a solution composed of exactly 1 mol of benzene and 1 mol of toluene.

Answer: (a) Since benzene may be mixed with toluene ideally without accompanying heat of solution, the free energy decrease associated with the mixing of 1 mol of benzene with toluene to bring the mole fraction of benzene to 0.5 will be

$$\Delta G_1 = \bar{G}_1 - G_1^\circ = -RT \ln 2$$

Conversely, the work required to extract 1 mol of benzene from an equimolal mixture is at least RT ln 2.

(b) The repetition of cycles of the separation of dn moles of benzene followed by further separation of dn moles of toluene results in the separation of 1 mol of benzene and toluene, respectively, with the same composition left in the solution. Thus, the minimum work necessary for each component would be RT ln 2. The total minimum work is thus 2RT ln 2.

PROBLEMS

2.1 The entropy of mixing in an ideal solution consisting of n_1 moles of the first component and n_2 moles of the second component may be expressed

$$\Delta S^M = -R\left(n_1 \ln \frac{n_1}{n_1 + n_2} + n_2 \ln \frac{n_2}{n_1 + n_2}\right)$$

Calculate the partial molal entropy of mixing, draw a diagram, and confirm the relation $\Delta S^M = n_1 \Delta\bar{S}_1 + n_2 \Delta\bar{S}_2$.

2.2 Assuming an ideal solution, calculate the relative activity of $C_2H_5OC_2H_5$ in the three-component solution containing equal weights of $C_2H_5OC_2H_5$, C_6H_6, and CS_2.

2.3 The chemical potential of the ith component of a two-component ideal solution is described over all temperature, pressure, and composition ranges, as follows:

$$\bar{G}_i = G_i^\circ(T, P) + RT \ln x_i$$

where G_i° is the chemical potential of the pure component i and x_i is the mole fraction. When such a solution is prepared by mixing the liquids, prove the following:

a. No change in volume is involved in mixing.

b. No absorption or emission of heat is involved in mixing.

c. The entropy increases by an amount equal to the entropy of ideal mixing.

d. No change of internal energy accompanies mixing.

2.4 Calculate the ratio of numbers of the possible arrangements among the two cases where N gaseous molecules are packed into a container of volumes V_1 and V_2, respectively. Then confirm that the Boltzmann equation

$$\Delta S = k \ln W$$

holds by comparing the ratio of the numbers, W, and the difference in the entropies of the two states:

$$\Delta S = Nk \ln \frac{V_2}{V_1}$$

chapter three

DILUTE SOLUTION

3.1 THE LAWS OF DILUTE SOLUTION

The several simple laws of dilute solution were discovered independently. However, these can be derived from one another by the application of thermodynamics. Let us first list these laws.

1. Henry's law (1803). The vapor pressure of the solute is proportional to its mole fraction in dilute solution [1].
2. Raoult's law (1887). The vapor pressure depression of the solvent is proportional to the mole fraction of solute in dilute solution [2].
3. Nernst's law (1891). The partition coefficient of a solute in two solvents in equilibrium at constant temperature is constant [3].
4. Van't Hoff's law (1887). As a result of the vapor pressure depression caused by dissolution of nonvolatile material in the solvent, the boiling point is elevated. The elevation of the boiling point is proportional to the concentration of solute. Similarly, the lowering of the freezing point is proportional to the concentration of solute [4].
5. Certain membranes, called semipermeable membranes, permit the free passage of solvent, but not of solute. If the pressure on the solution is increased to the point where the escaping tendency of the solvent is the same on both sides of the membrane,

equilibrium prevails, and solvent will pass through the membrane in neither direction. The osmotic pressure of the solution against the solvent is proportional to the concentration of the solute and is expressed by the following equation, due to van't Hoff [4]:

$$\pi V = n_2 RT \tag{3.1}$$

or

$$\frac{\pi}{C} = \frac{RT}{M_2} \tag{3.2}$$

where π, V, n_2, R, C, and M_2 are the osmotic pressure of the solution, the volume of the solution, the number of moles of solute in the solution, the gas constant, the concentration of solute (g/cm^3), and the molecular weight of the solute, respectively.

Let us explain these laws by considering a binary solution in which the mole fractions of solvent and solute are x_1 and x_2, respectively.

3.2 HENRY'S LAW

When the mole fraction of the solute approaches zero, its fugacity and partial pressure also approach zero. That is to say, in the neighborhood of the region where $x_2 = 0$ and $f_2 = 0$, from differential calculus, the following relations must hold:

$$\frac{\partial f_2}{\partial x_2} = \frac{f_2}{x_2} = \text{constant} \tag{3.3}$$

$$f_2 = kx_2 \tag{3.4}$$

As these relations hold for any continuous function, it is evident that the above relations always hold, regardless of the kind of solvents or whether the solvent is a pure substance or is itself a solution. The rule also holds for any component of a multicomponent system.

Mathematics alone cannot tell us how far into the range of finite concentration Henry's law, i.e., the law of dilute solution, may hold. Practically, it is almost valid in the range of composition where the solute molecules dissolve by molecular dispersion and are almost completely surrounded by solute molecules. As the concentration of solute increases, the solute molecules become, to some extent, closer to one another. However, the law still holds approximately in that range if the difference between solute and solvent is small. The constant of proportionality k depends on the properties of the solute and the solvent. In the case of ideal solutions $f_2(p_2)$ is proportional to x_2 and k is equal to the fugacity f_2° of the pure solute over the entire composition range. The most striking deviation from Henry's law is observed in solutions of surface-active substances above the saturation concentration of singly dispersed species. Here the aggregates of surface-active substances disperse to yield transparent solutions. (The solubility of the aggregates is large, approximately 20 to 50% so that workers in the field assumed that a ≈0.2% solution would obey the law of dilute solutions, creating much confusion and controversy. See Chap. 9.)

In an extremely dilute solution the mole fraction is proportional to the molal weight concentration and molal volume concentration. Thus the following two equations hold exactly:

$$f_2 = k_m m, \qquad f_2 = k_c C \tag{3.5}$$

where k_m and k_c are the experimentally determined constants. The original statement of Henry's law was that the vapor pressure of the solute is proportional to its concentration in dilute concentration, that is,

$$p_2 = k_c C \tag{3.6}$$

Typical experimental values of the Henry's law constant are given in Table 3.1.

In a solution of strong electrolyte, such as aqueous hydrochloric acid, the acid is almost completely dissociated into H^+ and Cl^- in dilute solution. Accordingly, the activities of H^+ and Cl^-

Table 3.1 Solubility of gases in water: Henry's Constant $(k \times 10^{-4})$[a]

Partial pressure	Henry's constant $(k \times 10^{-4})$		
	N_2 (19.4°C)	O_2 (23°C)	H_2 (23°C)
1.18	8.24	4.58	--
2.63	8.32	4.59	7.76
3.95	8.41	4.60	7.77
5.26	8.49	4.68	7.81
6.58	8.59	4.73	7.89
7.90	8.74	4.80	8.00
9.20	8.86	4.88	8.16

[a]Partial pressures of gases are expressed in atmospheres, and the composition of solute is expressed in mole fraction.

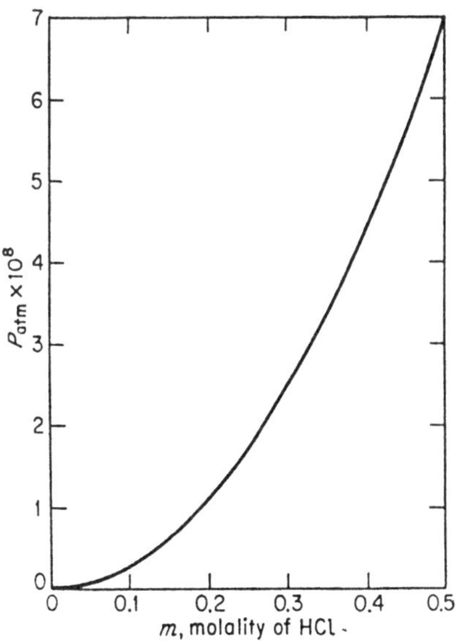

Fig. 3.1 Partial pressure of HCl in water at 25°C as a function of molality. From *Thermodynamics,* by G. N. Lewis and M. Randall, 2d ed. revised by K. S. Pitzer and L. Brewer. Copyright 1961. Used with permission of McGraw-Hill Book Company.

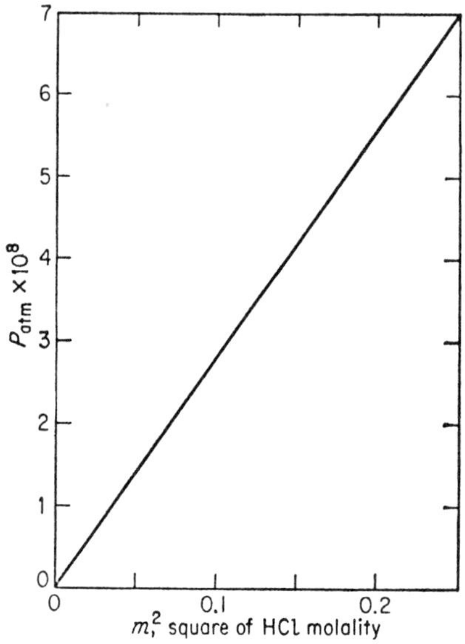

Fig. 3.2 Partial pressure of HCl in water at 25°C as a function of the square of the molality. From *Thermodynamics*, by G. N. Lewis and M. Randall, 2d ed. revised by K. S. Pitzer and L. Brewer. Copyright 1961. Used with permission of McGraw-Hill Book Company.

are nearly proportional to the stoichiometric concentration of the hydrochloric acid, and the activity of the hydrochloric acid is proportional to the square of its concentration. Therefore, as may be seen from Figs. 3.1 and 3.2, the partial pressure of hydrochloric acid is not proportional to the concentration of the solution but to its square.

3.3 RAOULT'S LAW

At constant temperature and pressure, the Gibbs-Duhem relation holds between the solvent and solute in binary solution:

$$x_1 \frac{\partial \ln f_1}{\partial x_1} = x_2 \frac{\partial \ln f_2}{\partial x_2} = \frac{x_2}{f_2} \frac{\partial f_2}{\partial x_2} \qquad (3.7)$$

Since Eq. (3.3) holds for an infinitely dilute solution, Eq. (3.7) is equal to unity. Therefore from the left-hand side of Eq.

(3.7) we obtain

$$d \ln f_1 = d \ln x_1 \tag{3.8}$$

Integrating and setting $f_1 = f_1^\circ$ at $x_1 = 1$, we obtain

$$f_1 = f_1^\circ x_1 = f_1^\circ(1 - x_2) \tag{3.9}$$

It is evident from the derivation of Eq. (3.9) that it applies regardless of the kinds of solutes, and that the fugacity of the solvent is proportional to the mole fraction; namely, the fugacity of the solvent in an infinitely dilute solution behaves as in an ideal solution. In the range where the vapor may be treated as an ideal gas, the vapor pressure, instead of the fugacity, may be used, and the following equation holds:

$$p_1 = p_1^\circ x_1 = p_1^\circ(1 - x_2) \tag{3.10}$$

This is Raoult's law. In the vapor pressure versus composition diagram of a binary solution, for example, the change of p_1 at $x_1 \approx 1$ in Fig. 4.5 clearly illustrates this relation. In the case when $x_1 \approx 1$, we derive the following relations from Eq. (3.8):

$$d \ln f_1 = dx_1 = -dx_2 \tag{3.11}$$

$$\left(\frac{\partial \ln f_1}{\partial x_2}\right)_{P,T} = -1 \tag{3.12}$$

3.4 PARTITION COEFFICIENT

If a small amount of ethyl alcohol is added to the two-phase mixture of benzene and water, the two phases form dilute solutions of alcohol. Accordingly, the fugacity of the alcohol is proportional to the mole fraction of alcohol in these two phases. In an equilibrium condition the fugacities of the alcohol in both phases are equal. Consequently, the ratio of the mole fractions of alcohol in both phases remains constant even when the amount of solute is changed. Thus,

$$f_2 = kx_2 = k'x_2' \tag{3.13}$$

Accordingly,

$$\frac{x_2'}{x_2} = \frac{k}{k'} = \text{constant} \tag{3.14}$$

This is called partition equilibrium.

3.5 LAWS OF BOILING-POINT ELEVATION AND FREEZING-POINT DEPRESSION

Since these two phenomena are similar, let us consider only the latter. Suppose ice and water exist in equilibrium in a thermally insulated vessel, and we then add a small amount of some solute. The fugacity of the water in the aqueous solution becomes smaller than the fugacity of the ice and some of the ice melts to form water. The temperature of the system is lowered thereby, and it reaches a new equilibrium when the chemical potential of the water molecules becomes equal to that of ice. This is expressed by the following relation:

$$\frac{\partial \ln f_1'}{\partial T}\, dT = \frac{\partial \ln f_1}{\partial T}\, dT + \frac{\partial \ln f_1}{\partial x_2}\, dx_2 \tag{3.15}$$

where f_1' and f_1 are the fugacities of ice and of the water of the aqueous solution, respectively. The former is a function of temperature only, while the latter is a function of temperature and of concentration. The change in fugacity due to the slight change of pressure may be neglected. Applying Eq. (1.46) to the above, we know that the left-hand side of that equation is equal to the heat of vaporization of ice divided by RT^2, and the first term of the right-hand side is equal to the heat of vaporization of the solvent divided by RT^2, while the value of the partial differential in the second term on the right is given by Eq. (3.12). Thus, we obtain

$$\frac{dT}{dx_2} = \frac{-RT^2}{\bar{H}_1 - H_1'} \tag{3.16}$$

where $\bar{H}_1 - H_1'$ is equal to the heat of fusion of the pure solvent. This is the equation for freezing-point depression in dilute solutions.

It is evident from Eq. (3.16) that the freezing-point depression is (1) proportional to the molal concentration of the solute but independent of the kind of solute and (2) inversely proportional to the heat of fusion of the solvent and proportional to the square of the freezing point.

Since the freezing point depression ΔT_f is also proportional to the molality in dilute solution, the following equation is equivalent to Eq. (3.16):

$$\Delta T_f = K_f m = K_f \frac{1000 w_2}{M_2 w_1} \tag{3.17}$$

where K_f is a constant for a particular solvent, and is called the molal freezing-point depression constant, and M_2 is the molecular weight of the solute. Melting points and molal freezing-point depression constants are shown for several major solvents in Table 3.2.

Similarly, the boiling-point elevation ΔT_b is proportional to the molality,

$$\Delta T_b = K_b m = K_b \frac{1000 w_2}{M_2 w_1} \tag{3.18}$$

where K_b is a constant for a particular solvent and is called the molal boiling-point elevation constant. Boiling points and boiling-point elevation constants for several solvents are given in Table 3.3.

Table 3.2 Molal Freezing-Point-Depression Constant

Solvents	T_f (°C)	K_f	ΔH_f (cal/g)
H_2O	0	1.86	79.7
C_6H_6	5.5	5.12	30.3
$C_6H_5NO_2$	5.7	8.1	22.5
CH_3COOH	16.7	3.9	44.7
Naphthalene	80	6.8	35.6
Camphor	178	40.0	10.7

Table 3.3 Molal Boiling-Point-Elevation Constant

Solvents	T_b (°C)	K_b	ΔH^V
$C_2H_5OC_2H_5$	34.4	2.02	89.3
CS_2	46.3	2.29	84.1
CH_2COCH_3	56.2	1.71	124.5
C_2H_3OH	78.3	1.22	204
C_6H_8	80.2	2.53	94.3
H_2O	100	0.52	539.5

Equation (3.16) was derived on the assumption that the solid phase separating on freezing was pure solid. This is the usual situation, but there are instances in which a solid solution is precipitated. Since the properties of the solid phase now depend on the composition as well as the temperature, the following equation must be used in place of Eq. (3.16):

$$\frac{\bar{H}_1 - \bar{H}_1'}{RT^2} dT = -(dx_2 - dx_2') \tag{3.19}$$

where x_2 and x_2' are the mole fractions of solute in the solution and in the solid solution, respectively.

If we denote the partition coefficient of the solute between solid phase and solution as k'/k, then $k'/k = dx_2'/dx_2$. Inserting this relation in (3.19):

$$\frac{dT}{dx_2} = \left(\frac{k'}{k} - 1\right)\frac{RT^2}{\bar{H}_1 - \bar{H}_1'} \tag{3.20}$$

Here $\bar{H}_1 - \bar{H}_1'$ is equal to the heat of fusion of the pure solvent in an infinitely dilute solution. When the solute does not dissolve in the solid phase, $k'/k = 0$. Thus, Eq. (3.20) becomes identical with Eq. (3.16). Although these relations hold in infinitely dilute solutions, thermodynamics alone cannot tell the concentration limits in which it is applicable. In the case of solid solution, deviation from ideality may be large in the solid compared to that in the liquid solution.

3.6 VAN'T HOFF'S LAW OF OSMOTIC PRESSURE

Membranes which permit the free passage of solvent molecules, but which do not permit the passage of solute molecules, are called semipermeable membranes. The phenomenon involving the diffusion of solvent into the solution, following the permeation through the membrane, is called osmosis.

In Fig. 3.3, A represents such a semipermeable membrane, separating an aqueous solution inside from pure water outside. When the water level inside and outside is equal, water will flow through the membrane from the outside to the inside because the fugacity of the water in the solution is smaller than that of pure water. As the pressure of the solution increases, due to the osmosis of the water, the system reaches an equilibrium (without further flow of the water) when the fugacity of the water in the solution reaches that of pure water. If p° represents the external pressure against the pure solvent and p represents the external pressure against the solution, at osmotic equilibrium, $p - p^\circ$ is the osmotic pressure.

In 1877, Pfeffer carried out osmotic pressure measurements, and the law governing the effect was derived by van't Hoff. The osmotic pressure must restore the fugacity of the solvent in the solution to that of the pure solvent. The relation between the

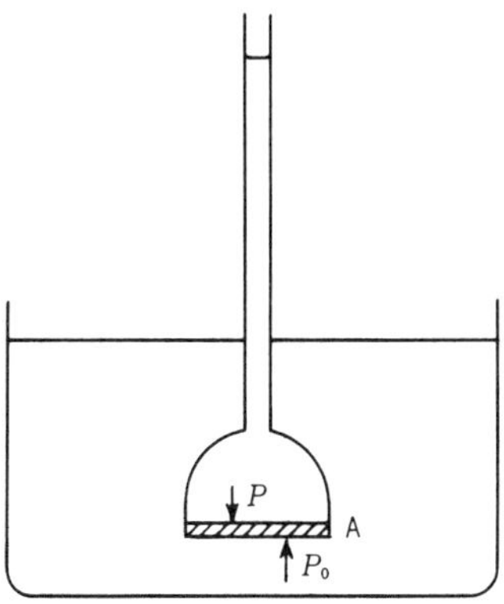

Fig. 3.3 Diagrammatic representation of osmosis.

fugacity of pure solvent and that in solution may be written as

$$d \ln f_1 = 0 = \frac{\partial \ln f_1}{\partial P} dp + \frac{\partial \ln f_1}{\partial x_2} dx_2 \tag{3.21}$$

Substituting Eqs. (1.45) and (3.12) into Eq. (3.21) we obtain

$$\frac{dp}{dx_2} = \frac{d(p - p^\circ)}{dx_2} = \frac{RT}{\bar{V}_1} \tag{3.22}$$

Now $p - p^\circ = 0$ when $x_2 = 0$, and, therefore, in dilute solution we obtain

$$\frac{d(p - p^\circ)}{dx_2} = \frac{p - p^\circ}{x_2} \tag{3.23}$$

$\bar{V}_1$ is equal to the molal volume of the pure solvent V_1 in dilute solution. Thus,

$$p - p^\circ = \frac{x_2 RT}{V_1} \tag{3.24}$$

Furthermore, since the number of moles of solute in 1 mol of solution is equal to the number of moles of solute in 1 mol of solvent (in the infinitely dilute solution), we obtain Eq. (3.1):

$$p - p^\circ = \frac{n_2 RT}{V} \tag{3.1}$$

where n_2 is the number of moles of solute in the volume V of solution, and where $p - p^\circ = \pi$.

The formal similarity of the equation to the perfect-gas equation is evident. Thus, the molecular weight of the solute may be obtained by the measurement of freezing-point depression or of osmotic pressure. Generally speaking, the freezing-point method is preferred for low-molecular-weight solutes because it provides a more precise measurement, while it is rather difficult to obtain a suitable semipermeable membrane for such a solute. The situation is reversed, however, for high-molecular-weight solutes and for polymer solutions, where the molal concentration of the solute is usually very low. Here, the osmotic pressure method provides the more sensitive measurement.

In the case of substances which dissociate into ions, if α represents the degree of dissociation, the osmotic pressure is enhanced by a quantity proportional to α. This coefficient $(1 + \alpha)$ is also known as the van't Hoff factor i.

3.7 EXAMINATION OF THE LAWS OF DILUTE SOLUTION

The thermodynamic functions which apply to binary solutions include H (or E), S, G (or F), P, T, V, $\Delta\bar{H}_1$, $\Delta\bar{H}_2$, $\Delta\bar{S}_1$, $\Delta\bar{S}_2$, $\Delta\bar{G}_1$ (or p_1), $\Delta\bar{G}_2$ (or p_2), $\Delta\bar{v}_1$, $\Delta\bar{v}_2$, x_1, and x_2. In order to be sure that all laws of dilute solutions are included, it is necessary to explore all possible relationships between these thermodynamic functions and the composition of the solution when x_2 is very small. Among these, the following combinations may be studied by direct experimental observations: (1) x_2 and p_2; (2) x_1 and p_1; (3) x and T; (4) x and p. Henry's law and the partition equilibrium law correspond to relation (1); Raoult's law corresponds to relation (2); freezing-point lowering and boiling-point elevation laws correspond to (3); and the osmotic pressure law corresponds to relation (4).

In dilute solution the change in extensive properties of the system on mixing for example, ΔH^M, ΔV^M are proportional to x_2. Except for the relations discussed above there is no relation which can be determined directly by experiment. However, we can derive relations between composition and partial molal quantities by applying our knowledge of the partial molal quantity. For instance, the combinations of x_i and $\Delta\bar{G}_i$, $\Delta\bar{S}_i$, $\Delta\bar{H}_i$, and $\Delta\bar{V}_i$ may be considered.

From Eq. (3.9):

$$\bar{G}_1 - G_1^\circ = \Delta\bar{G}_1 = RT \ln x_1 \tag{3.25}$$

$$-\left(\frac{\partial\Delta\bar{G}_1}{\partial T}\right)_p = \Delta\bar{S}_1 = -R \ln x_1 \tag{3.26}$$

$$\frac{\partial(\Delta\bar{G}_1/T)}{\partial(1/T)} = \Delta\bar{H}_1 = 0 \tag{3.27}$$

$$\left(\frac{\partial \Delta \bar{G}_1}{\partial P}\right)_T = \Delta \bar{V}_1 = 0 \tag{3.28}$$

are obtained as the properties of the solvent in dilute solutions. These equations show that the solvent behaves ideally; namely, there is no enthalpy or volume change, and the changes in entropy or free energy are ideal.

As for the solute, the following relations can be derived from Eq. (2.20):

$$\Delta \bar{G}_2 = RT \ln x_2 \gamma_2 \tag{3.29}$$

$$\Delta \bar{S}_2 = -R \ln x_2 - R \ln \gamma_2 - RT\left(\frac{\partial \ln \gamma_2}{\partial T}\right)_p \tag{3.30}$$

$$\Delta \bar{H}_2 = -RT^2\left(\frac{\partial \ln \gamma_2}{\partial T}\right)_p \tag{3.31}$$

$$\left(\frac{\partial \Delta \bar{G}_2}{\partial p}\right)_T = \Delta \bar{V}_2 = RT\left(\frac{\partial \ln \gamma_2}{\partial p}\right)_T \tag{3.32}$$

Equation (3.31) shows that the partial molal heat of dilution of the solute has a definite value even at infinite dilution, and Eq. (3.32) shows that the partial molal volume approaches a definite value, which can be different from the molal volume itself.

REFERENCES

1. W. Henry, Phil. Trans. Roy. Soc. (London), *29*, 274 (1803).
2. F. M. Raoult, Compt. Rend., *104*, 1430 (1887).
3. W. Nernst, Z. Physik. Chem., *8*, 110 (1891).
4. J. H. van't Hoff, Z. Physik. Chem., *1*, 481 (1887).

EXERCISES

3.1 Derive an expression for the change in solubility of a solid with temperature for the case in which the solubility is small. What happens if the solubility is not small enough?

Answer: Let f_2' represent the fugacity of the second component in the solid state, and f_2 its fugacity in a solution of mole fraction x_2. At saturation, maintaining equilibrium,

$$f_2' = f_2$$

$$d \ln f_2' = d \ln f_2$$

Thus,

$$\left(\frac{\partial \ln f_2'}{\partial T}\right)_p dT = \left(\frac{\partial \ln f_2}{\partial T}\right)_{p,x_2} dT + \left(\frac{\partial \ln f_2}{\partial x_2}\right)_{p,T} dx_2 \qquad (1)$$

In a dilute solution (below solubility limit)

$$\ln f_2 = \ln k + \ln x_2$$

$$\left(\frac{\partial \ln f_2}{\partial x_2}\right)_{p,T} = \frac{1}{x_2} \qquad (2)$$

Substituting Eq. (1.46) into Eq. (1):

$$\frac{H_2^* - H_2'}{RT^2} = \frac{H_2^* - \bar{H}_2}{RT^2} + \frac{1}{x_2}\frac{dx_2}{dT} \qquad (3)$$

$$\frac{d \ln x_2}{dT} = \frac{\bar{H}_2 - H_2'}{RT^2} \qquad (4)$$

These relations also hold for liquid solubility when the mutual solubility is small.

When the solubility is not small enough to justify the use of Eq. (2), the above does not hold precisely and may not be applied. Instead,

$$\left(\frac{\partial \ln f_2}{\partial \ln x_2}\right)_{p,T}\left(\frac{\partial \ln x_2}{\partial T}\right) = \frac{\bar{H}_2 - H_2'}{RT^2}$$

The calculation of the change of activity with concentration (the first term on the left-hand side of the above) is discussed in Sec. 6.2.

3.2 Show that the partial molal entropy of solution approaches an infinite positive value with increasing dilution.

Answer:

$$\bar{S}_i = -\left(\frac{\partial \bar{G}_i}{\partial T}\right)_p = -R \ln f_i + \text{constant}$$

$$= -R \ln x_i - R \ln f_i + \text{constant}$$

If

$$x_i \to 0, \qquad -\ln x_i \to \infty$$

Therefore

$$\lim_{x_i \to 0} S_i = +\infty$$

PROBLEMS

3.1 The vapor pressure p_2 of the solute in a binary dilute solution follows Henry's law. From this fact derive Raoult's law, and the laws of osmotic pressure, freezing-point lowering, and boiling-point elevation.

3.2 I_2 is dissolved in the phases CCl_4 and H_2O in contact with each other. When the concentration of I_2 in CCl_4 is 16.54 g/liter, the concentration in the aqueous phase is found to be 0.1934 g/liter. If 1 liter of aqueous solution containing 0.25 g I_2 is twice extracted with 10 cm^3 CCl_4, what percent of I_2 will remain in the aqueous phase?

3.3 When a substance is dissolved in water at 100°C, the vapor pressure is lowered 1 mmHg. Calculate the osmotic pressure of the solution.

3.4 The molal freezing-point-depression constant of C_6H_6 is 5.5. When 0.3251 g of naphthalene is dissolved in 30.90 g of C_6H_6 the melting point is depressed 0.454°C. Calculate the molecular weight of naphthalene.

chapter four

REGULAR SOLUTIONS

4.1 THE CONCEPT OF REGULAR SOLUTIONS

The term "regular solution" was first proposed in 1929 by Hildebrand to describe solutions which are characterized by the absence of any specific interactions between molecules, such as association, strong dipole interactions, and chemical effects, and in which the distributions and orientations are random [1].

In regular solutions the heat of mixing is taken into account when different kinds of molecules are mixed. Hildebrand developed his theory by assuming an almost completely random mixing arising from thermal energy which overcomes the difference in mutual molecular forces. When the differences between two kinds of molecules are large it may be supposed that there is no chance of random mixing. However, in this situation, the solubility is very small, and the system is so dilute that there is little contact between the solute molecules. Thus, the solution may be treated as though it were randomly mixed. In other words, Hildebrand assumed that the increase in entropy due to mixing is the same as in the case of athermal solutions. This assumption has been proved to be reasonably valid by various experiments. Guggenheim [2] has further applied the name "strictly regular solutions" to regular solutions composed of molecules of equal size.

Since the increase of entropy is now known not to be influenced

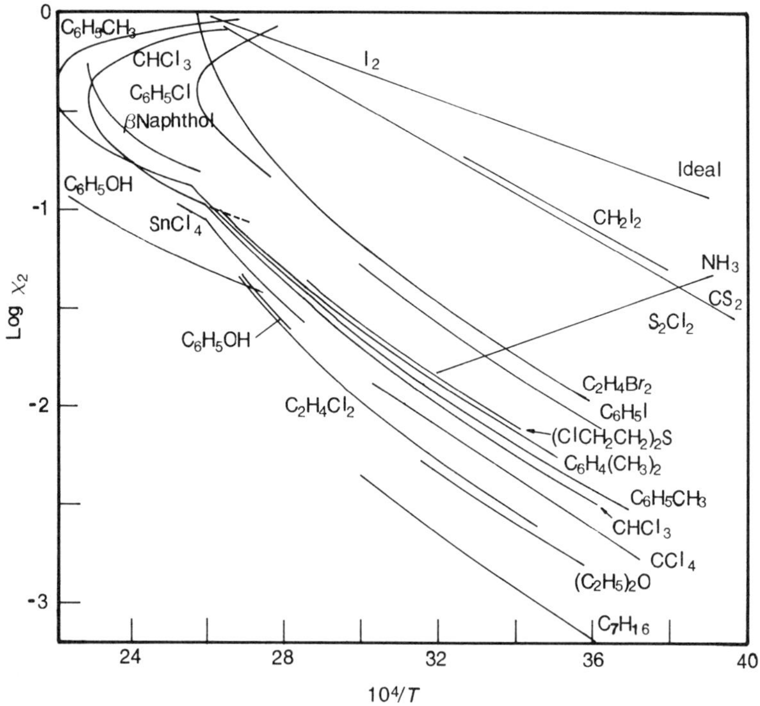

Fig. 4.1 Solubility of S_8. (Adapted from *Solubility of Nonelectrolytes*, by Hildebrand and Scott, 3d ed., 1950.)

strongly by molecular size (see Chap. 6), it appears that regular solution theory is sufficiently well established to warrant consideration both from the theoretical point of view and from the point of view of application to real systems. For most systems, the molecules are of different sizes and the heats of mixing per molecule are not equal. As the heat of solution of each component is proportional to the respective molal volumes in the regular solution theory, it can explain the partial molal heat of solution over all composition ranges.

The enthalpy term is derived by assuming that the intermolecular forces between two different molecules is given by the geometric mean of the respective molecular forces. In solutions which deviate somewhat from the regular solution equation, i.e., deviate from geometric mean, the heat term evaluated from vapor pressure or solubility may be used. Changes of solubility with temperature of a series of substances which form regular solutions are shown in Fig. 4.1, taking sulfur solutions as an example. In these regular solutions (except for the nonregular S_8-NH_3 system), the variation of solubility with temperature forms a family of curves [1].

4.2 ENTHALPY OF SOLUTION

Let the molal energies of vaporization of component 1 and component 2 be represented by ΔE_1^V and ΔE_2^V, respectively. Then the energy of condensation when n_1 moles of component 1 and n_2 moles of component 2 condense to a liquid is given by

$$-\Delta E_0 = -(n_1 \Delta E_1^V + n_2 \Delta E_2^V) \tag{4.1}$$

Since the components 1 and 2 in the gaseous state may be treated as ideal gases, the energy of the system does not change with mixing. In other words, the enthalpy and energy of gases change with temperature, but scarcely at all with pressure (or volume), because the mutual interaction between the molecules is extremely small. Let us calculate the energy of condensation when this mixture of gases is condensed into a $(n_1 + n_2)$ mole solution.

The volume of component 1 is n_1V_1 cm^3 and the fraction of the molecules of component 1 which are in contact with another molecule of the same species (1) is equal to the volume fraction of the component 1. Therefore, the energy of condensation due to the contacts of component 1 to component 1 is

$$\frac{-\Delta E_1^V}{V_1} n_1V_1 \frac{x_1V_1}{x_1V_1 + x_2V_2} = \frac{-\Delta E_1^V n_1x_1V_1}{x_1V_1 + x_2V_2} \tag{4.2}$$

A similar relation is also derived for component 2. The energy of condensation per unit volume for components 1 and 2 are $-\Delta E_1^V/V_1$ and $-\Delta E_2^V/V_2$, respectively. If we denote the condensation energy per unit volume due to the contacts of components 1 and 2 as $\Delta E_{12}^V/V_1^{1/2}V_2^{1/2}$, then the total condensation energy due to contact of component 1 with 2 is expressed [3,4]

$$-\frac{\Delta E_{12}^{v}}{V_1^{1/2}V_2^{1/2}}\left(n_1V_1\frac{x_2V_2}{x_1V_1 + x_2V_2} + n_2V_2\frac{x_1V_1}{x_1V_1 + x_2V_2}\right) \tag{4.3}$$

Accordingly, the total condensation energy of the solution due to the condensation of n_1 moles of 1 with n_2 moles of 2 is

$$-E_m = -\frac{\Delta E_1^{v} n_1^2 V_1 + 2\Delta E_{12}^{v} n_1 n_2 V_1^{1/2} V_2^{1/2} + \Delta E_2^{v} n_2^2 V_2}{n_1V_1 + n_2V_2} \tag{4.4}$$

The energy of mixing can thus be obtained from the difference between the condensation energy for the solution and for the pure liquids, by subtracting Eq. (4.1) from Eq. (4.4):

$$\Delta E_m = \frac{n_1V_1n_2V_2}{n_1V_1 + n_2V_2}\left(\frac{\Delta E_1^{v}}{V_1} - \frac{2\Delta E_{12}^{v}}{V_1^{1/2}V_2^{1/2}} + \frac{\Delta E_2^{v}}{V_2}\right) \tag{4.5}$$

If the intermolecular force is solely due to London dispersion forces, the intermolecular pair potential between the two molecules is

$$|u_{12}| = \frac{2(I_1I_2)^{1/2}}{I_1 + I_2}(|u_1||u_2|)^{1/2} \tag{1.8}$$

It is evident from Eq. (1.8) that in addition to the geometric mean of the molecular forces, the potential is also a function of the ionization potentials of the respective molecules.

Ionization potentials of some typical liquids are summarized in Table 4.1; as can be seen, the differences among them are not great. Hence,

$$\frac{2(I_1I_2)^{1/2}}{I_1 + I_2} \approx 1$$

For example, when $I_1 = 12$ and $I_2 = 9$, the ionization energy term is equal to 0.99. Thus, in regular solution theory, the

Table 4.1 Ionization Potential, I

Compounds	I (eV)	Compounds	I (eV)
$n\text{-}C_7F_{16}$	12.1	$n\text{-}C_7H_{16}$	10.4
$CCl_2F\cdot CClF_2$	12.0	C_2H_6	12.8
CCl_4	11.5	C_6H_6	9.2
$CHCl_3$	11.4	$C_6H_5CH_3$	8.8
CH_2Cl_2	11.3	$p\text{-}(CH_3)_2C_6H_4$	8.6
CH_3Cl	10.7	$1,3,5\text{-}(CH_3)_3C_6H_3$	8.4
CH_3CHCl_2	11.1	C_6H_5I	8.7
CH_2ClCH_2Cl	11.1	CS_2	10.4
C_2H_5Cl	11.0	Cl_2	13.2
CH_2Br_2	10.5	Br_2	12.8
C_2H_5Br	10.3	I_2	9.7
C_2H_5I	9.3	H_2O	12.6
CH_3I	9.1	NH_3	11.2

intermolecular potential can be approximated by the geometric mean of the intermolecular forces between the two molecules.

If ΔE_{12}^{V} can be expressed as the geometric mean of ΔE_1^{V} and ΔE_2^{V}, Eq. (4.5) becomes

$$\Delta E_m = \frac{n_1V_1n_2V_2}{n_1V_1 + n_2V_2}\left[\left(\frac{\Delta E_1^{V}}{V_1}\right)^{1/2} - \left(\frac{\Delta E_2^{V}}{V_2}\right)^{1/2}\right]^2 \tag{4.6}$$

In regular solution theory, the term $(\Delta E^{V}/V)^{1/2}$ appears with great frequency. It is a suitable parameter for describing intermolecular forces and solubility. It is called the "solubility parameter" and is given the symbol δ. Hence,

$$\Delta E_m = \frac{n_1V_1n_2V_2}{n_1V_1 + n_2V_2}(\delta_1 - \delta_2)^2 \tag{4.7}$$

The partial molal energy of solution of component 2 can be obtained from Eq. (4.7) by partial differentation with respect to n_2.

$$\left(\frac{\partial \Delta E_m}{\partial n_2}\right)_{n_1} = \Delta \bar{E}_2 = V_2 \phi_1{}^2 (\delta_1 - \delta_2)^2 \tag{4.8}$$

where ϕ_1 is the volume fraction of component 1.

4.3 THE SOLUBILITY EQUATION

The entropy and energy (enthalpy) of solution of component 2 in the pure liquid state into component 1 were derived in the previous section. The change of volume due to the mixing is usually small, so that the change in energy may be equated with the change in enthalpy (a more refined treatment is given in Chap. 6).

If we express the change in Gibbs energy of component 2 due to transfer into solution from the pure liquid state to solute by ΔG_2, we obtain

$$\Delta \bar{G}_2 \approx RT \ln x_2 + V_2 \phi_1{}^2 (\delta_1 - \delta_2)^2 \tag{4.9}$$

If the fugacity of component 2 in the pure liquid state is represented by f_2° and that in the solution by f_2, the following relation is obtained:

$$RT \ln \frac{f_2}{f_2^\circ} = RT \ln a_2 \approx RT \ln x_2 + V_2 \phi_1{}^2 (\delta_1 - \delta_2)^2 \tag{4.10}$$

This very important equation is called the basic solubility equation.

When the vapor pressure is small, the ratio of the fugacities is equal to the ratio of the vapor pressures. The partial vapor pressure of component 2 in equilibrium with the solution is thus, from Eq. (4.10),

$$p_2 = p_2^\circ x_2 \exp \frac{V_2 \phi_1{}^2 (\delta_1 - \delta_2)^2}{RT} \tag{4.11}$$

The exponential term, $\exp[V_2 \phi_1{}^2 (\delta_1 - \delta_2)^2/RT]$, varies only slightly with a small change in x_2 in dilute solution, i.e., when $x_2 \approx 0$ and $\phi_1 \approx 1 = \text{const}$. Hence, the vapor pressure p_2 is proportional to x_2, i.e., Henry's law holds. Comparing Eq. (4.11) with Eq. (3.4), it is seen that the physical meaning of the constant k in the empirical equation is

$$k = p_2^{\circ} \exp \frac{V_2(\delta_1 - \delta_2)^2}{RT} \tag{4.12}$$

In the case where $x_2 \approx 1$ and $\phi_1 \approx 0$, we obtain

$$p_2 \approx p_2^{\circ} x_2 = p_2^{\circ}(1 - x_1) \tag{4.13}$$

i.e., the solubility equation also satisfies Raoult's law.

Substituting Eq. (4.10) into Eq. (2.20), the activity coefficient is given by

$$\gamma_2 = \exp \frac{V_2 \phi_1^{\,2} (\delta_1 - \delta_2)^2}{RT} \tag{4.14}$$

Since, for a given solution, $V_1(\delta_1 - \delta_2)^2$ is constant, the change in activity coefficient with concentration depends solely on the volume fraction of solvent ϕ_1:

$$\ln \gamma_2 = \frac{B\phi_1^{\,2}}{RT} \tag{4.15}$$

When the molecular sizes of components 1 and 2 are equal, $\phi_1 = x_1$. Therefore,

$$\ln \gamma_2 = \frac{Bx_1^{\,2}}{RT} \tag{4.16}$$

The change of activity with composition is shown in Fig. 4.2, where the activities are plotted assuming B = 0, 1.15RT, 2RT, and 2.88RT, respectively, and where the molecular sizes are equal.

For the case where B = 2RT, the activity-composition curve has a point of inflection at $x_1 = x_2 = 0.5$. When the heat of mixing is smaller than 2RT, the two liquids dissolve in each other over the entire composition range. For large heat of mixing (B > 2RT), however, the two liquids separate. The compositions of the two

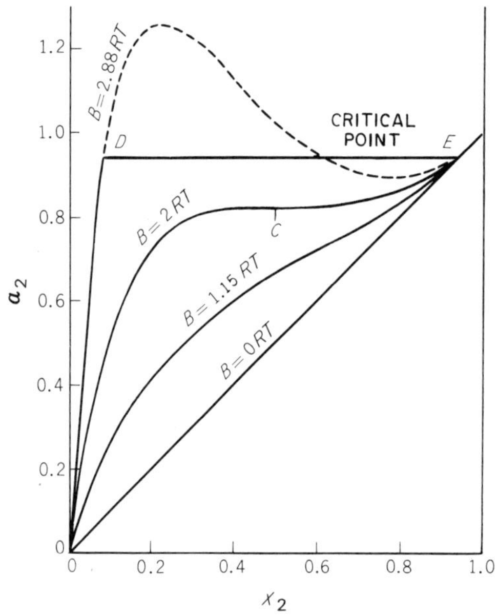

Fig. 4.2 Relative activity versus mole fraction.

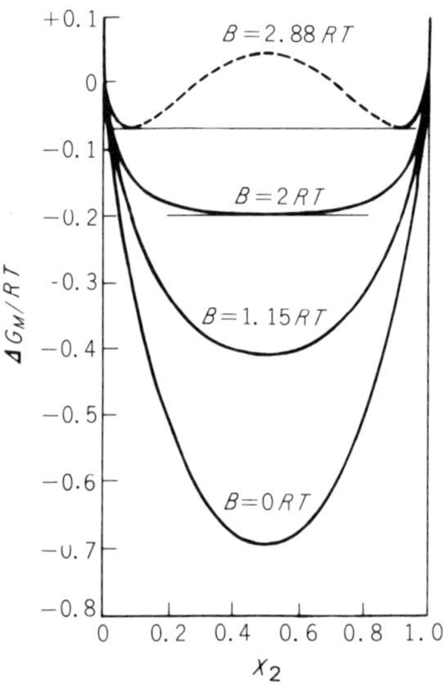

Fig. 4.3 Gibbs free energy of mixing. Broken curve for B = 2.88 indicates phase separation.

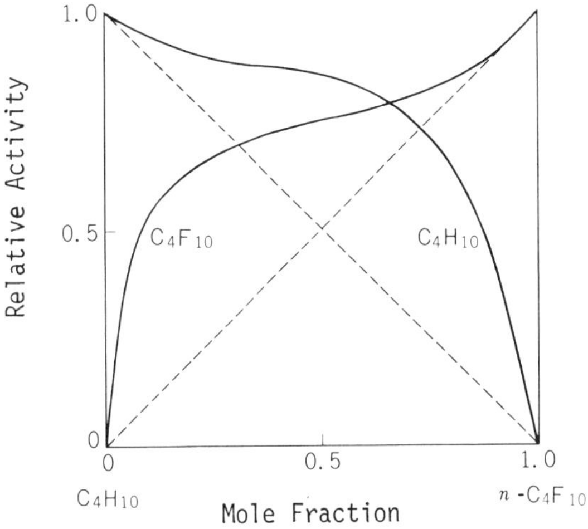

Fig. 4.4 Relative activities in solutions of C_4F_{10}-C_4H_{10}.

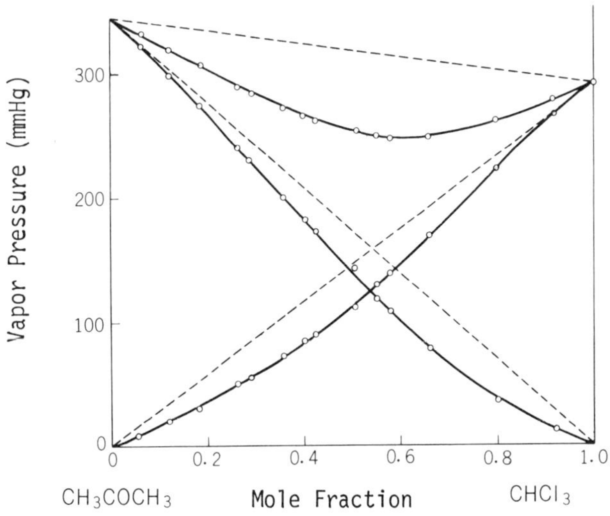

Fig. 4.5 Partial and total vapor pressures in CH_3COCH_3-$CHCl_3$ solutions at 35.17°C.

coexisting phases are shown as D and E in Fig. 4.2. The molal Gibbs free energy of mixing is shown in Fig. 4.3. When B = 2.88RT, an unstable region appears, shown as a broken curve. Examples of the change of activity and vapor pressure of a binary solution are shown in Figs. 4.4 and 4.5.

In regular solutions γ_2 should be larger than unity because the heat of mixing is always positive. However, the theory may also be applied where γ_2 is less than one, using the empirical negative heat of mixing.

4.4 EVALUATION OF SOLUBILITY PARAMETERS

It is clear from Eq. (4.10) that the solubility can be determined when the activity of the solute, the molal volumes, temperature, and solubility parameters of solute and solvent are known. Thus, it is necessary to know the value of δ for the liquids under study. As the definition $\delta = (\Delta E^V/V)^{1/2}$ implies, δ changes gradually with temperature, hence it must be calculated at the temperature for which solubility is required. The units of δ are $(\text{cal/cm}^3)^{1/2}$.

4.4.1 Calculation from Heat of Vaporization

The molal volumes and specific gravities at various temperatures are generally known. The evaluation of δ is mainly dependent on the determination of ΔE^V. If the vapor pressure of the liquid is not large, the vapor can be treated as an ideal gas. Thus, since $\Delta H^V - RT = \Delta E^V$, we may write

$$\delta = \left(\frac{\Delta H^V - RT}{V}\right)^{1/2} \tag{4.17}$$

where ΔH^V is the heat of vaporization. When the heat of vaporization is not known at the desired temperature, the Clapeyron-Clausius equation

$$\left(\frac{\partial \ln p}{\partial T}\right)_{\text{sat}} = \frac{\Delta H^V}{RT^2}$$

may be used to calculate the heat of vaporization from the change in vapor pressure with temperature.

4.4.2 *Estimation by the Hildebrand Rule*

For substances where neither the heat of evaporation nor change of vapor pressure with temperature is accurately known, the heat of vaporization may be estimated by the aid of the Hildebrand rule, provided the liquid is nonpolar or nearly so [5]. This rule states that the entropy of vaporization per mole is the same for all normal liquids if measured at the temperature at which their vapors have equal volumes. This leads to the conclusion that the entropy and heat of vaporization at any temperature T is a function of only two parameters, the temperature T and the normal boiling point T_b. An empirical curve is obtained when the heats of vaporization of various substances at 25°C are plotted against their boiling points (Fig. 4.6). The heat of vaporization of all normal liquids fall on or near this curve.

The following empirical equation may be used to estimate the heat of vaporization at 25°C (298 K):

$$\begin{aligned} \Delta H^V_{298} &= -2950 + 23.7T_b + 0.020T_b^2 \\ \Delta E^V_{298} &= -3542 + 23.7T_b + 0.020T_b^2 \end{aligned} \tag{4.18}$$

where the units are calories per mole.

4.4.3 *Estimation from Surface Tension*

Since surface tension arises as a result of intermolecular forces, it may be expected to be related to the solubility parameter. Hildebrand proposed the following empirical relation [1]:

$$\delta = 4.1\left(\frac{\gamma}{V^{1/3}}\right)^{0.43} \tag{4.19}$$

The constant term is equal to 4.1 at 25°C and varies with temperature. Hildebrand reported that it was restricted to simple

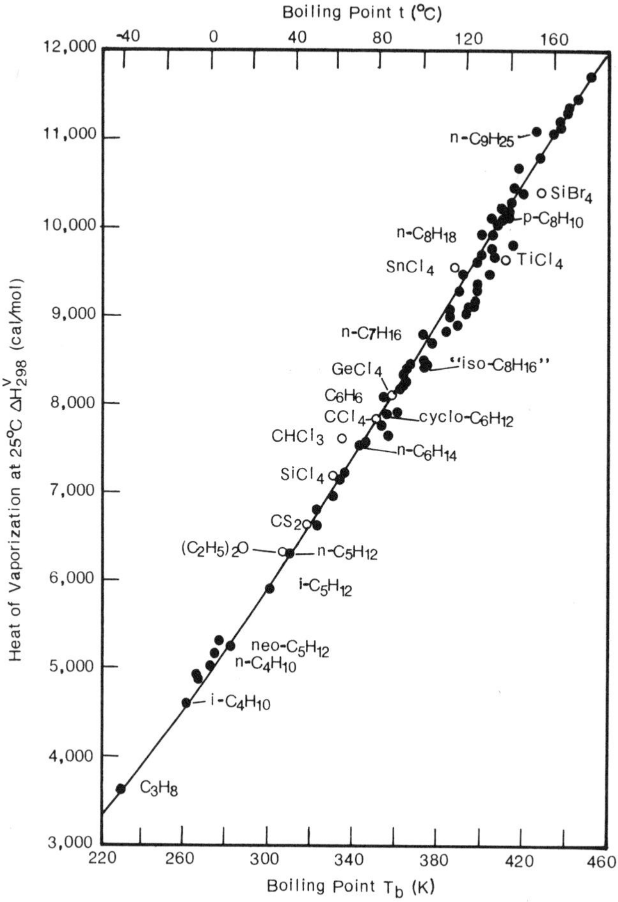

Fig. 4.6 Heats of vaporization at 25°C. (Adapted from *Solubility of Nonelectrolytes*, by Hildebrand and Scott, 3d ed., 1950.)

molecules, without polar groups.

However, Beerbower [6] has demonstrated recently that in the form

$$\delta = 3.741\left(\frac{\gamma}{V^{1/3}}\right)^{0.5} \tag{4.20}$$

the relation applies to a large number of liquids, including many excluded by Hildebrand. Equation (4.20) has the additional advantage over Eq. (4.19) of being homogeneous in the exponent. Becher [7] has shown recently that Eq. (4.20) can be derived either on geometric grounds or on the basis of a quasi-thermodynamic argument by considering the solubility parameter to be equal to the surface energy of the molecule per unit volume.

In addition to these methods, one may calculate δ from the internal pressure or critical constants of the liquid. However, these are not especially useful, since for molecules where these quantities are known, one also usually knows the heat of vaporization, vapor pressure, and boiling point.

4.4.4 Estimation from Solubility Data

Finally, of course, one may determine the solubility parameter in an empirical way by substituting solubility data for gases and solids in solvents of known solubility parameter into Eq. (4.10) and solving for the unknown solubility parameter. For example, the solubility parameter of iodine, as calculated from its solubility in various solvents, is summarized in Table 4.2. The deviation of the respective δ values is usually ± 0.2.

Table 4.2 Solubility Parameter of Iodine Evaluated from Solubility in Various Solvents

Solvents	δ_1	$\delta_2(I_2)$
CS_2	10.0	14.1
$CHCl_3$	9.3	14.3
$TiCl_4$	9.0	14.1
CCl_4	8.6	14.2
cyclo-C_6H_{12}	8.2	14.0
n-C_7H_{16}	7.5	13.5
$SiCl_4$	7.6	14.1
$CCl_2F \cdot CClF_2$	7.5	14.3
n-C_7F_{16}	6.0	14.4
Average		14.1

Table 4.3 Solubility Parameters at 25°C

Compounds	V_{298} (cm^3)	ΔH^V_{298} (kcal)	δ
Br_2	51	7.34	11.5
I_2	59	--	14.1
S_8	135	--	12.7
P_4	70	12.6	13.1
OsO_4	58	9.80	12.6
MoF_6	84	6.36	8.3
WF_6	88	6.25	8.0
UF_6	96	7.2	8.9
$Si(CH_3)_4$	136	5.80	6.2
$TiCl_4$	111	9.62	9.0
$SiCl_4$	115	7.19	7.6
$SiBr_4$	127	10.38	8.8
$GeCl_4$	115	8.09	8.1
$SnCl_4$	118	9.55	8.7
SnI_4	151	19.63	11.7
CCl_4	97	7.83	8.6
$CHCl_3$	81	7.41	9.2
CH_2Cl_2	64	6.84	9.8
$CHBr_3$	88	10.3	10.5
CH_3I	63	6.7	9.9
CH_2I_2	81	--	11.8
C_2H_5Cl	74	5.7	8.3
C_2H_5Br	75	6.5	8.9
C_2H_5I	81	7.7	9.4
1,2-$C_2H_4Cl_2$	85	7.7	9.1
1,2-$C_2H_4Br_2$	90	9.9	10.2
1,1,1-$C_2H_3Cl_3$	100	7.8	8.5
cis-$C_2H_2Cl_2$	76	6.9	9.1
trans-$C_2H_2Cl_2$	78	6.8	9.0
C_2Cl_4	103	9.5	9.3

Table 4.3 (continued)

Compounds	V_{298} (cm^3)	ΔH^V_{298} (kcal)	δ
CS_2	61	6.7	10.0
$C_2H_5OC_2H_5$	105	6.36	7.4
H_2O	18	--	23.8
n-C_5H_{12}	116	6.40	7.1
$(CH_3)_2CHC_2H_5$	117	6.03	6.8
$(CH_3)_4C$	122	5.35	6.2
n-C_6H_{14}	132	7.57	7.3
n-C_7H_{16}	148	8.75	7.4
n-C_8H_{18}	164	9.92	7.5
iso-C_8H_{18}	166	8.40	6.9
n-$C_{16}H_{34}$	294	19.38	8.0
cyclo-C_5H_{10}	95	6.85	8.1
cyclo-C_6H_{12}	109	7.91	8.2
$CH_3C_6H_{11}$	128	8.46	7.8
cis-Decalin	154	12.5	8.8
trans-Decalin	159	12.1	8.6
1-C_6H_{12}	126	7.34	7.3
1-C_8H_{16}	158	9.70	7.6
C_6H_5	89	8.10	9.2
$CH_3C_6H_5$	107	9.08	8.9
$C_2H_5C_6H_5$	123	10.10	8.8
o-$(CH_3)_2C_6H_4$	121	10.38	9.0
m-$(CH_3)_2C_6H_4$	123	10.20	8.8
p-$(CH_3)_2C_6H_4$	124	10.13	8.8
1,3,5-$(CH_3)_3C_6H_3$	140	11.35	8.8
Styrene	116	10.5	9.3
Naphthalene	123		9.9
Anthracene	(150)		9.9
Phenanthrene	158		9.8
n-C_6F_{14}	205	7.75	5.9

Table 4.3 (continued)

Compounds	V_{298} (cm^3)	ΔH^V_{298} (kcal)	δ
$n\text{-}C_7F_{16}$	226	8.69	6.0
cyclo-C_6F_{12}	170	6.9	6.1
$CF_3C_6F_{11}$	196	7.9	6.1
$(C_4F_9)_3N$	360	13.0	5.9
cyclo-$C_4Cl_2F_6$	142	--	7.1
$CCl_2F{\cdot}CClF_2$	120	7.9	7.5
$C_7F_{15}H$	215	9.01	6.3

Solubility parameters at 25°C for a number of substances are listed in Table 4.3.

Since the solubility parameter is defined as $\delta = (\Delta E^V/V)^{1/2}$, it is to be expected that the value of δ will be smaller for the lower homologues of the hydrocarbons (whose density is low). However the solubilities of a given solute, such as iodine, in a series of saturated hydrocarbons does not vary as much as would be expected from the spread in solubility parameters, e.g., 6.9 for iso-C_8H_{18} and 8.0 for n-$C_{16}H_{34}$. Actually, the change of solubility of solute in a homologous series of solvents is not large. Hence, one may assume that the solubility of particular solute in various members of a homologous series would not be greatly different.

For substances whose intermolecular forces consist mostly of dispersion forces, i.e., in regular and related solutions, the heat of mixing may be roughly evaluated by assuming the geometric mean for the intermolecular forces between the different kinds of molecules. However, in polar molecules, the energy of the system is decreased because of the orientation of dipolar molecules, and thus the estimation of the heat of solutions derived from the solubility parameter differences may not be accurate.

On the other hand, it has been suggested recently that the polar interactions can be taken care of by the assumption that the solubility parameter can be considered as being made up of various components, that is,

$$\delta^2 = \delta_d^{\ 2} + \delta_p^{\ 2} + \delta_h^{\ 2} \tag{4.21}$$

where δ_d is the contribution due to dispersion forces, δ_p is the contribution due to dipole-dipole interactions, and δ_h is the contribution due to hydrogen bonding. Hansen and Beerbower [8] have described how such "three-dimensional" solubility parameters may be used.

As the regular solution equation is derived on the basis of a counterbalance of the entropy and energy of mixing, without any specific interaction, it explains the relation between the energy of mixing and the solubility of fused salts, solution of semiconductors, and so on, as well as ordinary solutions. On the other hand, we should not (mathematically) expect the theory to explain specific phenomena, since the regular solution equation is not a function of specific interactions.

REFERENCES

1. J. H. Hildebrand, J. Am. Chem. Soc., *51*, 66 (1929); J. H. Hildebrand and R. L. Scott, *Regular Solutions*, Prentice-Hall, New York, 1962; J. H. Hildebrand, J. M. Prausnitz, and R. L. Scott, *Regular and Related Solutions*, Van Nostrand Reinhold, New York, 1970.
2. E. A. Guggenheim, Proc. Roy. Soc. A, *148*, 304 (1935).
3. G. Scatchard, Chem. Rev., *8*, 321 (1931).
4. G. Scatchard, J. Am. Chem. Soc., *56*, 995 (1934).
5. J. H. Hildebrand, J. Am. Chem. Soc., *37*, 970 (1915); *40*, 45 (1918).
6. A. Beerbower, J. Colloid Interface Sci., *35*, 46 (1971).
7. P. Becher, J. Colloid Interface Sci., *38*, 291 (1972).
8. C. Hansen and A. Beerbower, *Kirk-Othmer Encyclopedia of Chemical Technology*, 2d ed., Supplement Volume, Wiley-Interscience, New York, 1971, pp. 889-910.

EXERCISES

4.1 The activity of component 1 in a binary solution is given by the equation $R \ln a_1 = R \ln x_1 + Ax_2^2 + Bx_2^3$, where x_1 and x_2 are the respective mole fractions, and A and B are constants. The equation is valid over the entire concentration range from pure liquid 1 to pure liquid 2. Derive an expression for the activity of component 2.

Answer: Differentiating the given equation:

$$d \ln a_1 = \frac{dx_1}{x_1} + \frac{2Ax_2\, dx_2}{R} + \frac{3Bx_2^2\, dx_2}{R}$$

Applying the Gibbs-Duhem equation, $x_1\, d \ln a_1 + x_2\, d \ln a_2 = 0$,

$$d \ln a_2 = -\frac{dx_1}{x_2} - \frac{2Ax_1\, dx_2}{R} - \frac{3Bx_1x_2\, dx_2}{R}$$

$$= \frac{dx_2}{x_2} + \frac{2Ax_1\, dx_1}{R} + \frac{3Bx_1(1 - x_1)dx_1}{R}$$

$$(\ln a_2)_1^{x_2} = (\ln x_2)_1^{x_2} + \left(\frac{Ax_1^2}{R}\right)_0^{x_1} + \left[\frac{3B}{R}\left(\frac{x_1^2}{2} - \frac{x_1^3}{3}\right)\right]_0^{x_1}$$

$$R \ln a_2 = R \ln x_2 + Ax_1^2 + B\left(\frac{3}{2} x_1^2 - x_1^3\right)$$

4.2 At the maximum or minimum for the curve for B = 2.88RT in Fig. 4.2, prove that the curve corresponding to the other component becomes minimum or maximum for the same composition.

Answer: From the Gibbs-Duhem equation for a binary system

$$x_1 \frac{\partial \ln f_1}{\partial x_1} + x_2 \frac{\partial \ln f_2}{\partial x_1} = 0$$

Rewriting the equation, it becomes

$$\frac{x_1}{f_1} \frac{\partial f_1}{\partial x_1} + \frac{x_2}{f_2} \frac{\partial f_2}{\partial x_1} = 0$$

For a maximum or minimum

$$\frac{\partial f_1}{\partial x_1} = 0$$

and accordingly

$$\frac{\partial f_2}{\partial x_1} = 0$$

Furthermore, when $\partial f_1/\partial x_1 > 0$, $\partial f_2/\partial x_1 < 0$; hence the curve of component 2 becomes a minimum when that of 1 is a maximum, and vice versa.

PROBLEMS

4.1 The energy of mixing of a solution consisting of n_1 moles of liquid 1 and n_2 moles of liquid 2 is given by

$$\Delta E_m = \frac{n_1 V_1 n_2 V_2}{n_1 V_1 + n_2 V_2}(\delta_1 - \delta_2)^2$$

Derive the partial molal energy of mixing for component 2.

4.2 The vapor pressure of bromine, dissolved in carbon tetrachloride, is $p_2/x_2 = 0.539$ atm in the range $x_2 = 0$ to 0.025. The vapor pressure of pure bromine is 0.280 atm at the same temperature. Calculate the activity coefficient of bromine in dilute solution in carbon tetrachloride. Also calculate the relative activity at $x_2 = 0.5$.

4.3 Calculate the solubility parameter of carbon tetrachloride from the information that its density at 25°C is 1.59 and its boiling point is 76.8°C.

chapter five

ESTIMATION OF THE SOLUBILITY OF GASES, LIQUIDS, AND SOLIDS IN LIQUIDS

5.1 ESTIMATION OF GAS SOLUBILITY

In chemistry and chemical engineering, it is frequently desirable to be able to estimate the solubility of gases in liquids at various temperatures and pressures. There are usually insufficient good experimental data available, and these are usually obtained at about 25°C. One is thus required to estimate gas solubilities from little or essentially no information.

As the logarithms of the saturated vapor pressure versus the reciprocal absolute temperature gives an almost straight line, the saturated vapor pressure p_2° of a gas at a desired temperature can be obtained with little difficulty. If p_2 represents the pressure of the gas, the relative activity of the gas is given by $a_2 = p_2/p_2^\circ$. From Eq. (4.10) we obtain

$$\ln \frac{p_2^\circ}{p_2^\circ} \approx \ln \frac{f_2}{f_2^\circ} \approx \ln x_2 + \frac{V_2\phi_1{}^2(\delta_1 - \delta_2)^2}{RT} \tag{5.1}$$

The solubility of x_2 may be calculated by inserting the appropriate values of p_2/p_2°, V_2, δ_1, δ_2, and T in Eq. (5.1).

In many cases, gas solubilities are very small, so that the condition $\phi_1 \approx 1$ holds. It is evident from Eq. (5.1) that the solubility of a gas is nearly proportional to its vapor pressure (Henry's law, see Sec. 3.2). The ratio of the fugacities, rather than the ratio of the vapor pressures, should be used in order to obtain the solubility more precisely, because the saturated vapor pressure of the gas p_2° is usually large, and the temperature at which the solubility is sought is in many cases higher than the critical temperature of the gas.

In order to use the fugacities, it becomes necessary to estimate the value of f_2°, i.e., the fugacity of the gas when it exists in the hypothetical superheated liquid. This may be done by the use of the principle of corresponding states. The ratios of fugacities of various liquids divided by their critical pressures versus reduced temperatures are plotted in Fig. 5.1 [1].

As discussed in Chap. 1 (Sec. 1.2), the principle of corresponding states amounts to the statement that the equation of all substances can be expressed by a single reduced equation, when there is no strong association or dissociation, provided the reduced property is plotted on the reduced scale. Thus, the properties of almost all substances fall on one curve. Accordingly, f_2° at any temperature may be evaluated from Fig. 5.1, if T_c and P_c of the gas are known.

If the vapor pressure p_2 of the gas (solute) is not too high, we may assume $p_2 = f_2$, without loss of accuracy. We must also estimate the solubility parameter δ_2 and the molal volume V_2 of the gas (solute) in the liquid. Assuming that neither V_2 nor δ_2 changes significantly with temperature, the values derived from solubility data at 25°C may generally be used. In Table 5.1 values at V_2 and δ_2 at 25°C are summarized for twelve gases.

Thus, the solubility of the gas at a pressure p_2 can be estimated by substituting f_2°, V_2, δ_1, δ_2, and the temperature in Eq. (5.1). However, it is not possible to estimate the solubility from Eq. (5.1) when the solubility is affected by a specific interaction between solute and solvent. In addition, if the gas pressure is large the deviation of p_2 from f_2 becomes appreciable, and the following correction is necessary.

The relation between the fugacity and the total pressure P is given by

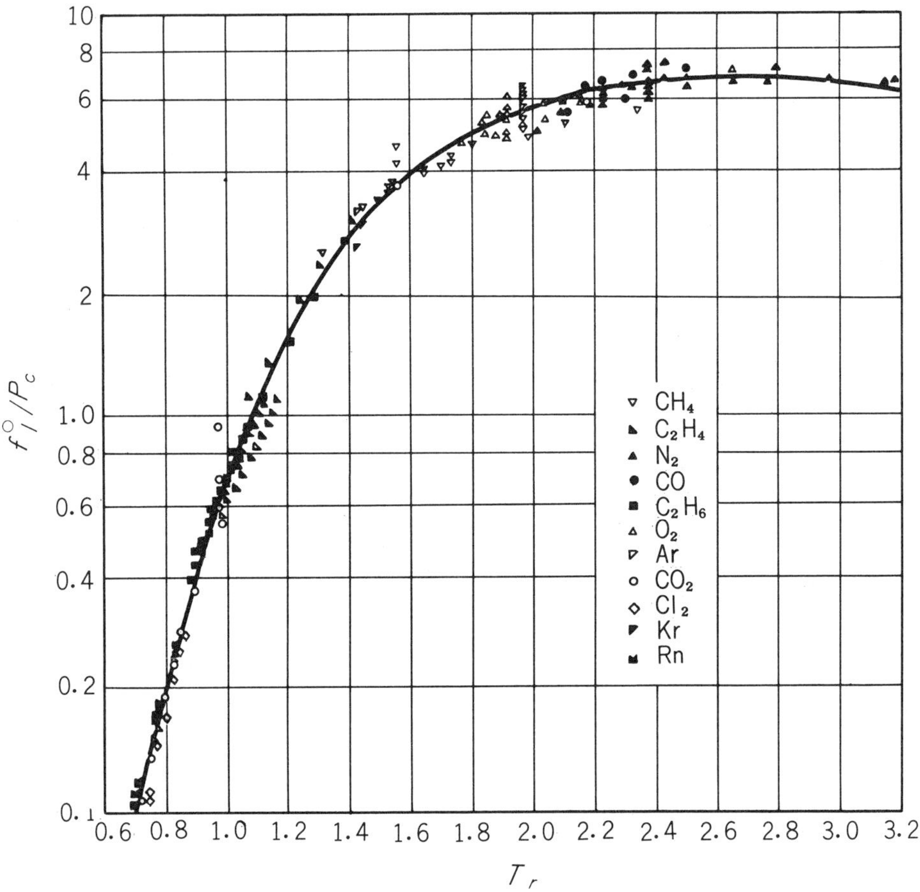

Fig. 5.1 Fugacity of hypothetical liquid as a function of reduced temperature. From Ref. 1. Reprinted with permission. Copyright by the American Chemical Society.

$$\left(\frac{\partial \ln f}{\partial P}\right)_V = \frac{V}{RT} \tag{1.45}$$

Accordingly, the ratio of the fugacity to the pressure may be written

$$\frac{\partial \ln(f/p)}{\partial P} = \frac{\partial \ln f}{\partial P} - \frac{\partial \ln p}{\partial P} = \frac{V}{RT} - \frac{1}{P} \tag{5.2}$$

Table 5.1 Molal Volumes (Liquid) and Solubility Parameters for Gases at 25°C

Gases	V_2 (cm^3)	δ_2 [(cal/cm^3)$^{1/2}$]	δ_2
N_2	32.4	2.59	5.9 (77°K)
CO	32.1	3.13	6.1 (82°K)
O_2	33.0	4.00	7.2 (90°K)
Ar	57.1	5.33	7.0 (87°K)
CH_4	52.0	5.68	6.8 (112°K)
CO_2	55.0	6.00	8.9 (223°K)
Kr	65.0	6.40	7.5 (121°K)
C_2H_4	65.0	6.60	7.9 (169°K)
C_2H_6	70.0	6.60	7.6 (184°K)
Rn	70.0	6.83	8.4 (211°K)
Cl_2	74.0	8.7	9.8 (239°K)
H_2	35	--	2.5 (20°K)

Equation (5.2) may be integrated from zero pressure (at which p = f) to the pressure p

$$\ln \frac{f}{p} = \int_0^p \left(\frac{V}{RT} - \frac{1}{P} \right) dp \tag{5.3}$$

If the virial coefficients of the gas in the following equation are known

$$PV = RT + BP + CP^2 + DP^3 + \cdots \tag{5.4}$$

substitution of Eq. (5.4) into (5.3) yields

$$\ln \frac{f}{p} = \int_0^p \left(\frac{B}{RT} + \frac{CP}{RT} + \frac{DP^2}{RT} + \cdots \right) dp \tag{5.5}$$

or

$$\ln \frac{f}{p} = \frac{BP}{RT} + \frac{CP^2}{RT} + \frac{DP^3}{RT} + \cdots \tag{5.6}$$

From this equation, the fugacity f may be obtained at any pressure p.

When the virial coefficients B, C, D, ... of a gas are not exactly known, the equation of state proposed by Berthelot in 1907 is useful:

$$PV = RT + \frac{9RPT_c}{128P_c}\left(1 - 6\frac{T_c^2}{T^2}\right) \tag{5.7}$$

Equation (5.7) may be regarded as a virial equation which contains only the second virial coefficient. Substituting Berthelot's equation into Eq. (5.3), we obtain for the fugacity of a gas

$$\ln\frac{f}{p} = \frac{9T_cP}{128TP_c}\left(1 - 6\frac{T_c^2}{T^2}\right) \tag{5.8}$$

The right-hand side of Eq. (5.8) is a reduced quantity and therefore the relation holds for any gas. If the deviation of f from p is small, i.e., $f/p \approx 1$, we obtain

$$\frac{f}{p} \approx 1 + \frac{9T_cP}{128TP_c}\left(1 - 6\frac{T_c^2}{T^2}\right) \tag{5.9}$$

The fugacity f_2 is obtained by substituting the reduced pressure and temperature in Eq. (5.8) or (5.9), and the ratio f_2/f_2°, i.e., the relative activity of the gas at relatively high temperatures and pressures may be calculated with fair accuracy [2].

The solubilities of various gases at 25°C and one atmosphere pressure are plotted against the solubility parameters of the solvents in Fig. 5.2 [3]. If the solubility parameter of a solvent is known, the solubility of gas may be estimated from Fig. 5.2. The points for CO_2 in CCl_4, $C_6H_5CH_3$, and C_6H_6 lie well above the line for nonbasic solvents, indicating electron donor-acceptor interaction. In Fig. 5.3, the gas solubilities are plotted against the energy of vaporization at the respective boiling points as well as the force constants for the gases, ε/k, where ε is the minimum value of the intermolecular pair potentials and k is the Boltzmann

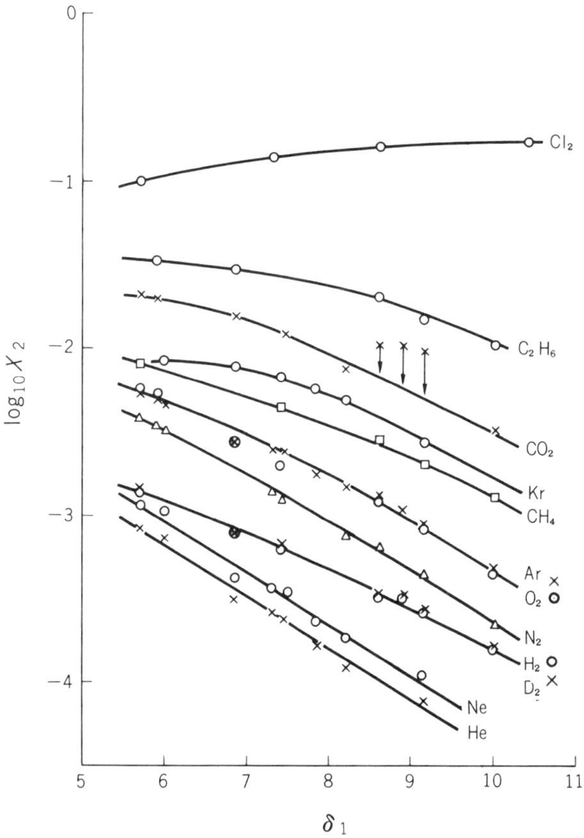

Fig. 5.2 Gas solubilities versus solubility parameters of solvents (25°C, 1 atm). From Ref. 3. Reprinted with permission. Copyright by the American Chemical Society.

constant. It is possible to estimate unknown gas solubilities from these empirical correlations [3]. While the values of ε/k scatter, depending on the source of the data, the molal energies of vaporization at their normal boiling points, where different gases are approximately in corresponding states, do not. We may use ε/k for gases whose energies of vaporization are not known, or determine consistent values of ε/k from the ε/k versus ΔE_V^V relation [4].

The entropy of solution of a gas may be calculated from the change of solubility with temperature, using the thermodynamic

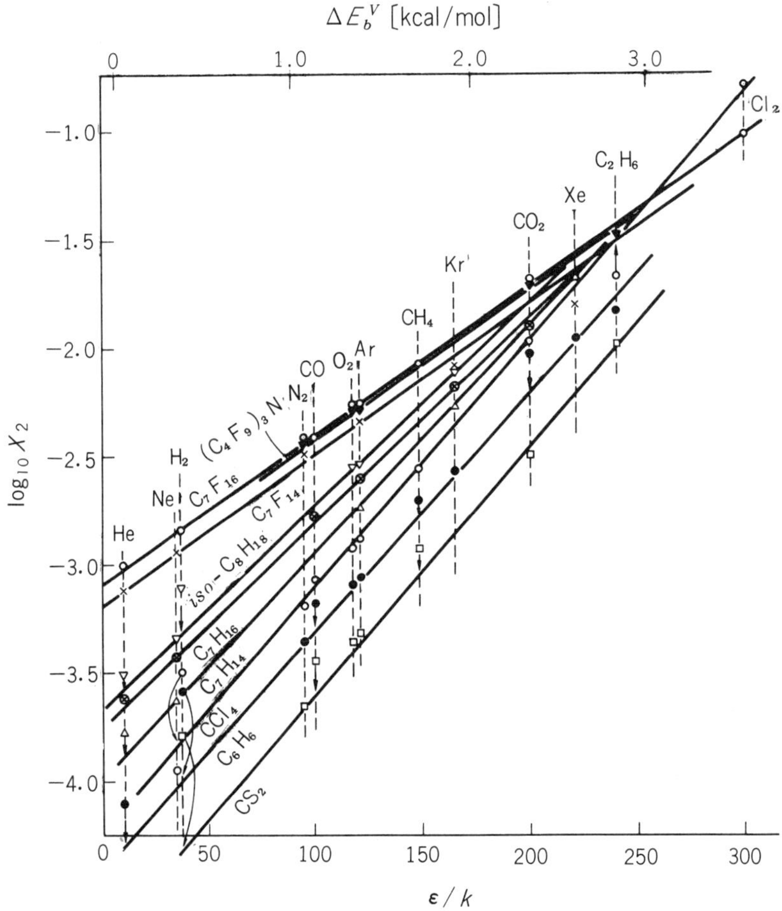

Fig. 5.3 Solubility as a function of the force constant of gases (25°C, 1 atm) and their energy of vaporization at their respective boiling points. Reprinted with permission from J. Phys. Chem., *65*, 331 (1961). Copyright by the American Chemical Society.

relation Eq. (5.12) [4]. Applying Eq. (2.14) to the second component in the solute and gaseous states,

$$\bar{S}_2 - S_2^g = -\left[\frac{\partial(\bar{G}_2 - G_2^g)}{\partial T}\right]_{p,x}$$

$$= \left[\frac{\partial(\bar{G}_2 - G_2^g)}{\partial \ln x_2}\right]_{p,T}\left(\frac{\partial \ln x_2}{\partial T}\right)_{(\bar{G}_2 - G_2^g),p} \tag{5.10}$$

In the first factor on the right-hand side of Eq. (5.10) $\bar{G}_2 - G_2^g = RT \ln f_2/f_2^g$, f_2 is the fugacity of the solute in the solution, and f_2^g is that of the gaseous solute. The fugacity f_2^g is not a function of x_2. We may write

$$\left[\frac{\partial(\bar{G}_2 - G_2^g)}{\partial \ln x_2}\right]_{p,T} = RT\left[\frac{\partial \ln f_2}{\partial \ln x_2}\right]_{p,T}$$

$$= RT\left[\frac{\partial \ln a_2}{\partial \ln x_2}\right]_{p,T} \tag{5.11}$$

Substituting Eq. (5.11) into (5.10), we obtain

$$\bar{S}_2 - S_2^g = R\left(\frac{\partial \ln a_2}{\partial \ln x_2}\right)_{p,T}\left(\frac{\partial \ln x_2}{\partial \ln T}\right)_{sat,p} \tag{5.12}$$

where the arbitrary standard state of the gas is taken as one atmosphere. When the pressure is low (and, accordingly, the solubility is low) the following equation holds:

$$\left(\frac{\partial \ln a_2}{\partial \ln x_2}\right)_{p,T} \approx 1$$

The entropy of solution is thus directly related to the change of solubility with temperature. The entropy of solution of various gases in seven typical solvents at 25°C is plotted against ideal entropy of mixing (as $-R \ln x_2$) in Fig. 5.4.

The following features are to be noted:

1. The solubility parameters of the six solvents cover a wide

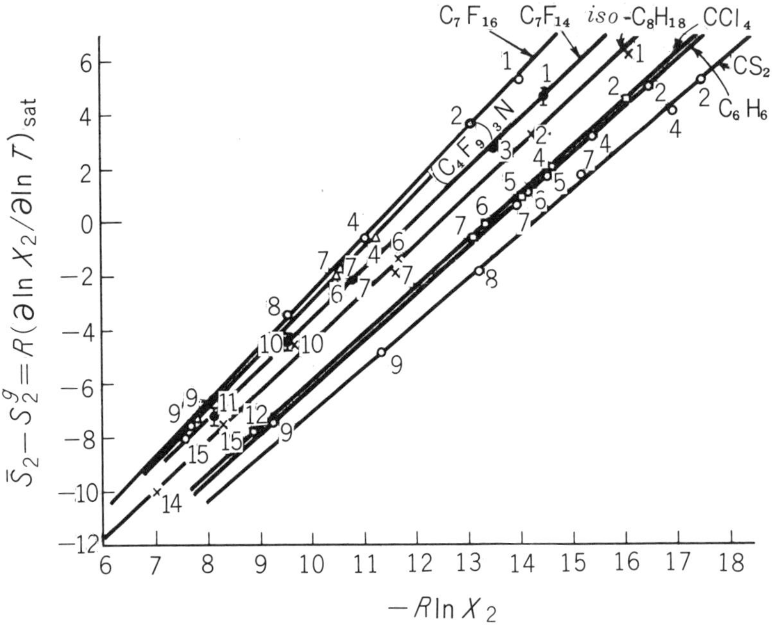

Fig. 5.4 Relation between the entropy of solution of gases and their solubility (25°C, 1 atm). 1, He; 2, H_2; 3, Ne; 4, N_2; 5, CO; 6, O_2; 7, Ar; 8, CH_4; 9, CO_2; 10, Kr; 11, Xe; 12, C_2H_2; 13, C_2H_4; 14, C_2H_6; 15, SF_6. Reprinted with permission from J. Phys. Chem., *65*, 331 (1961). Copyright by the American Chemical Society.

range. Therefore, it may be concluded that the relation between the solubility and its change with temperature falls into a certain domain.

2. The solubilities at temperatures other than 25°C can be calculated from the solubility at that temperature and the rate of change of solubility with temperature. The dividing line between positive and negative temperature coefficients is at $\bar{S}_2 - S_2^g = 0$, and it is possible to predict the temperature dependence from the solubility at 25°C, using Fig. 5.4.

Generally speaking, the entropy of solution becomes more negative in passing from the less soluble to the more soluble gases, i.e., gases which are more like the solvent in solubility parameter. When $-R \ln x_2 = 0$, $\bar{S}_2 - S_2^\circ$ represents the entropy of condensation of the solvent itself from a hypothetical pressure of 1 atm to liquid. The entropy of vaporization of cyclohexane calculated from liquid to 1 atm vapor pressure at 25°C is equal to 21.0 cal/deg mol, exactly the value of the intercept of the line for solutions of gases in cyclohexane [4] (see Fig. 6.3). The solubilities of gases generally decrease (or increase a little) with increase in temperature because the relative activity of gases at a definite pressure decreases with increase of temperature.

5.2 ESTIMATION OF MUTUAL SOLUBILITY OF LIQUIDS

If p_2 represents the vapor pressure of component 2 in the gas phase, x_2 and x_2' the mole fractions of the component 2 in the liquid phase 1 and 2, and ϕ_1 and ϕ_1' the volume fractions of component 1 in the liquid phases 1 and 2, the following relation may be obtained from the solubility equation, (4.10):

$$\ln a_2 = \ln \frac{p_2}{p_2^\circ} = \ln x_2 + \frac{V_2\phi_1^{\,2}(\delta_1 - \delta_2)^2}{RT}$$

$$= \ln x_2' + \frac{V_2\phi_1'^{\,2}(\delta_1 - \delta_2)^2}{RT} \qquad (5.13)$$

In the case that the solubility of component 1 in component 2 is small, substituting $\phi_1' \approx 0$ and $x_2' \approx 1$ into Eq. (5.13) we obtain

$$\ln a_2 \approx 0 \approx \ln x_2 + \frac{V_2\phi_1^{\,2}(\delta_1 - \delta_2)^2}{RT} \qquad (5.14)$$

The solubility may be estimated from this equation. When the solubility is appreciable the solution activity of component 2 may be calculated from $\ln a_2 = \ln x_2'$ (Raoult's law), where x_2' is the mole fraction of component 2 (solvent) in liquid 2. The solubility of component 1, x_1', can be estimated from an equation similar to Eq. (5.14).

It is evident from Eq. (5.14) that the solubilities of solutes of similar solubility parameters in a given solvent depends on V_2, and that it rapidly decreases as V_2 increases at constant temperature.

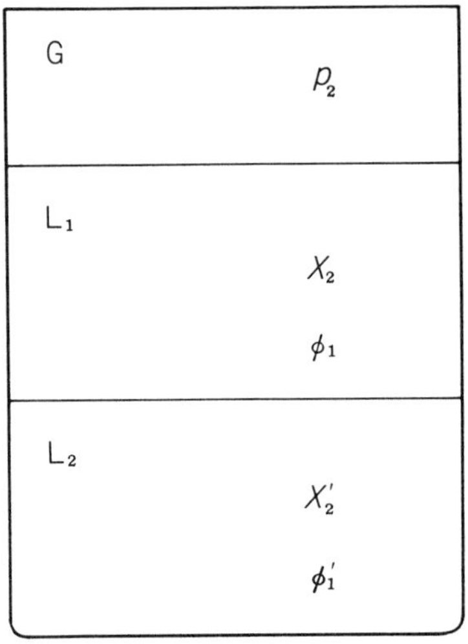

Fig. 5.5 Mutual solubility of liquids (see text).

The solubilities of a homologous series of hydrocarbons in water correspond to this case. On the other hand, when a given solute is dissolved in similar solvents of varying molal volumes, the solubility does not change substantially, because the heat of solution is almost constant. The solution of water in various hydrocarbons corresponds to this situation. Since the solubility is expressed in terms of mole fraction, the amount dissolved in a definite volume when the solubility is constant decreases if the molal volume of the solvent increases.

These theoretical predictions are borne out by the following experimental results:

1. *Solubility of water in hydrocarbons.* The solubility of water in hydrocarbons is discussed here in order to show that regular solution theory can be applied to the solubility of such an extraordinary liquid as water, if the other component is normal, and to demonstrate the situation in which the difference in solubility parameters of the two components is almost constant. The apparent solubility parameter of water is quite large, $\delta \approx 23.8$, owing to the strong hydrogen bonding of water and to its small molar

volume. Accordingly, the difference between the solubility parameters of water and of hydrocarbons becomes almost constant. Hence, the solubility of water in hydrocarbons is found to be of the order of 10^{-3} to 10^{-4} in units of mole fraction as indicated in Table 5.2 [5].

It may be concluded from Table 5.2 that

1. The longer the length of the hydrocarbon chain, the higher will be the δ-value, and the solubility of water will increase.
2. The solubility of water is not influenced strongly by the degree of branching of the hydrocarbon chain.
3. The presence of one double bond increases the solubility by a factor of approximately seven.

Table 5.2 Solubility of Water in Hydrocarbons, x_2, at 20°C [5]

Hydrocarbons	$x_2 \times 10^4$	δ_1	δ_2 (H_2O)
n-C_4H_{10}	2.1	6.7	24.1
n-C_5H_{12}	3.6	7.05	24.1
n-C_6H_{14}	5.3	7.3	24.0
n-C_7H_{16}	7.3	7.45	24.0
n-C_8H_{18}	9.0	7.55	24.0
iso-C_4H_{10}	2.4	6.25	23.3
iso-C_5H_{12}	3.6	6.75	23.8
cyclo-C_6H_{12}	4.65	8.2	24.8
C_6H_6	19.0	9.15	24.2
$C_6H_5CH_3$	22.0	8.9	24.0
1-C_4H_8	12.4	6.7	22.3
2-C_4H_8	14.0	7.0	22.8
1-C_7H_{14}	57.0	7.7	22.0
1,3-C_4H_6	19.0	7.6	22.7
1,5-C_6H_{10}	46.0	8.2	22.6

These observations are based on the solubility at 25°C. Abnormal behavior is evident, however, if we consider the heat of solution. When water is dissolved in hydrocarbons the hydrogen bonds between the water molecules are broken. Thus, the change of solubility with temperature becomes abnormally large, resulting in a rapid decrease in solubility as the temperature is depressed.

2. *Solubility of paraffin-chain alcohols in water.* The regular solution concept can be applied to the solubilities of homologous series of compounds in water. For example, the logarithm of the solubility of paraffin-chain alcohols in water decreases linearly with the hydrocarbon chain length, as shown in Table 5.3. This relation also can be understood by reference to Eq. (5.14). As the energy of solution increases in arithmetic progression with the increase in hydrocarbon chain length, the solubility decreases geometrically. The logarithm of the solubility versus the number of carbon atoms in alcohols varies linearly, as shown in Fig. 5.6 [6]. As neither the phenomena nor the theory are specific, similar relations hold for the other homologous series of compounds in water [7]. More precisely, the enthalpy and entropy of solution are not as large as would be expected from Eq. (4.10), and the small solubility change with temperature, i.e., the small apparent heat of solution, regardless of the small solubility may be explained by

Table 5.3 Solubility of n-Aliphatic Alcohols in Water at 25°C

Compounds	Solubility (moles/liter)
C_4H_9OH	0.97
$C_5H_{11}OH$	0.25
$C_6H_{13}OH$	0.059
$C_7H_{15}OH$	0.0146
$C_8H_{17}OH$	0.0038
$C_9H_{19}OH$	0.00097
$C_{10}H_{21}OH$	0.000234

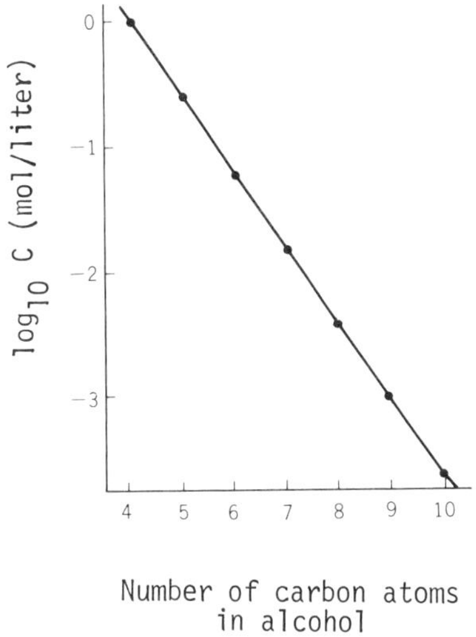

Fig. 5.6 Relation between the solubility of alcohols in water and the number of carbon atoms in the alcohol. From Ref. 6. Reprinted with permission.

the formation of an "iceberg" of water molecules around the hydrocarbon chains (see Chap. 7).

5.3 CRITICAL SOLUTION TEMPERATURE AND COMPOSITION OF LIQUID-LIQUID MIXTURES

Owing to the decrease of the heat of mixing in liquid-liquid mixtures, the mutual solubility of the liquids increases with temperature, finally yielding a homogeneous phase. An example of the change in mutual solubility of liquids with temperature is shown in Fig. 5.7 [8].

As shown in Fig. 4.2, the activity (vapor pressure) versus composition curve has a point of inflection at x = 1/2 if the heat of mixing is 2RT. That is to say, the change of activity with composition has neither a minimum nor a maximum value at the critical point, and thus the conditions $(\partial a/\partial x)_{p,T} = 0$ and $(\partial^2 a/\partial x^2)_{p,T} = 0$ both hold. Applying these conditions to Eq. (4.10), the critical temperature and composition are obtained. The first differentiation

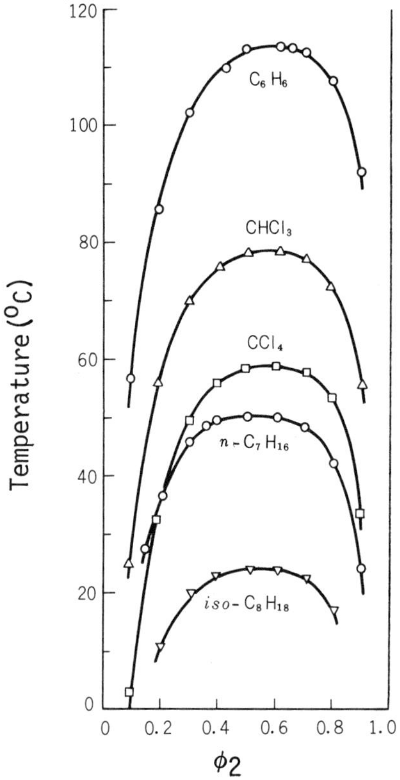

Fig. 5.7 Liquid-liquid solubility (ϕ_2) of n-C_7F_{16} in various liquids.

yields

$$RT_c = \frac{2\,x_{1c}x_{2c}V_1^{\,2}V_2^{\,2}(\delta_1 - \delta_2)^2}{(x_{1c}V_1 + x_{2c}V_2)^3} \tag{5.15}$$

and the second differentiation

$$x_{1c} = \frac{(V_1^{\,2} - V_1V_2 + V_2^{\,2})^{1/2} - V_1}{V_2 - V_1} = 1 - x_{2c} \tag{5.16}$$

and furthermore

$$\left(\frac{x_{1c}}{x_{2c}}\right) = \frac{(V_1^2 - V_1V_2 + V_2^2)^{1/2} - V_1}{V_2 - (V_1^2 - V_1V_2 + V_2^2)^{1/2}} \tag{5.17}$$

If the molecular volumes are not too different, the relation $x_1V_1 \approx x_2V_2$ holds, as shown in Fig. 5.7 (note that $x_1V_1 = x_2V_2$ implies $\phi = 0.5$). Substituting this relation into Eq. (5.15), we obtain

$$RT_c \approx \frac{V_1V_2}{V_1 + V_2}(\delta_1 - \delta_2)^2 \tag{5.18}$$

Equations (5.15) to (5.18) are approximate relations because the system is not a completely random mixture near the critical temperature. Empirically, Eq. (5.17) holds better than the others.

It is worthwhile to note that the critical composition is a function only of the molal volume. The critical composition can be estimated from Eq. (5.17), provided that the molal volumes are known. Also, $\delta_1 - \delta_2$ can be estimated from Eq. (5.15) if the critical temperature is determined, and vice versa. If the entropy of mixing is expressed by Eq. (2.18), the activity (or solubility) is given as follows:

$$\ln a_2 = \ln \phi_2 + \phi_1\left(1 - \frac{V_2}{V_1}\right) + \frac{V_2\phi_1^2(\delta_1 - \delta_2)^2}{RT} \tag{5.19}$$

Applying the conditions

$$\left(\frac{\partial \ln a_2}{\partial x_2}\right)_{p,T} = 0, \qquad \left(\frac{\partial^2 \ln a_2}{\partial \ln x_2^2}\right)_{p,T} = 0 \tag{5.20}$$

to Eq. (5.19), the critical temperature and composition may be derived:

$$RT_c = \frac{2V_1V_2(\delta_1 - \delta_2)^2}{(V_1^{1/2} + V_2^{1/2})^2} \tag{5.21}$$

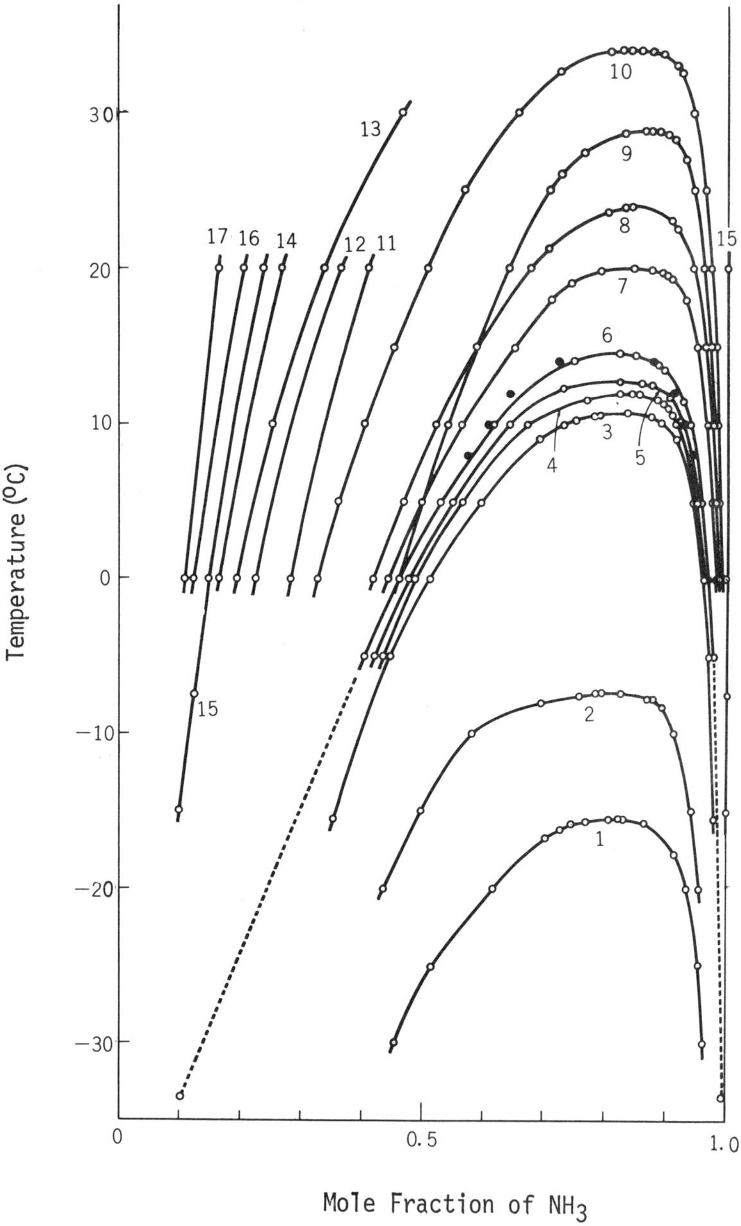

Fig. 5.8 Liquid-liquid solubility of ammonia in various solvents. 1, styrene; 2, $C_6H_5CH_3$; 3, $C_6H_5C_2H_5$; 4, 5, 6, o-, p-, m-$(CH_3)_2C_6H_5$; 7, iso-$C_3H_7C_6H_5$; 8, n-$C_3H_7C_6H_5$; 9, 1-$CH_3C_{10}H_7$; 10, 1,3,5-$(CH_3)_3C_6H_5$; 11, tetralin; 12, C_8H_{16}; 13, cyclo-C_6H_{10}; 14, iso-C_8H_{18}; 15, n-C_7H_{16}; 16, $CH_3C_6H_{11}$; 17, decalin. From Ref. 9. Reprinted with permission.

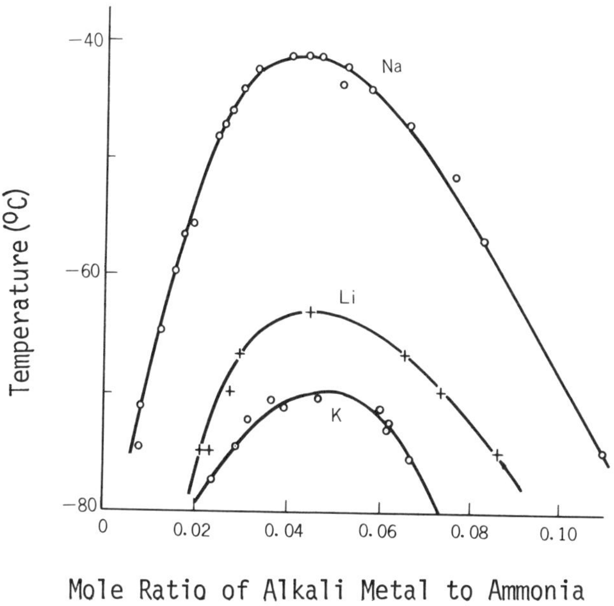

Fig. 5.9 Phase equilibria for the system ammonia-alkali metal. From Ref. 10. Reprinted with permission. Copyright by the American Chemical Society.

$$\frac{x_{2c}}{x_{1c}} = \left(\frac{V_1}{V_2}\right)^{3/2} \tag{5.22}$$

In contrast to the strong hydrogen bond and dipole-dipole interactions between water molecules, the weaker interactions between ammonia molecules, resulting in a solubility parameter of 13, permit it to mix well with hydrocarbons. Ammonia has a rather high dielectric constant (22.4 at -33.4°C), large dipole moment (1.49 Ds), and a small molal volume (26.55 cm^3 at 0°C). It is a basic solvent. These properties make it a remarkably versatile solvent. It mixes with water and is even a solvent for alkali metals.

The mutual solubilities of ammonia in organic liquids [9] and with alkali metals [10], respectively, are shown in Figs. 5.8 and 5.9, respectively.

In ordinary liquid-liquid solution, the mutual solubility increases with increasing temperature, and a single homogeneous phase

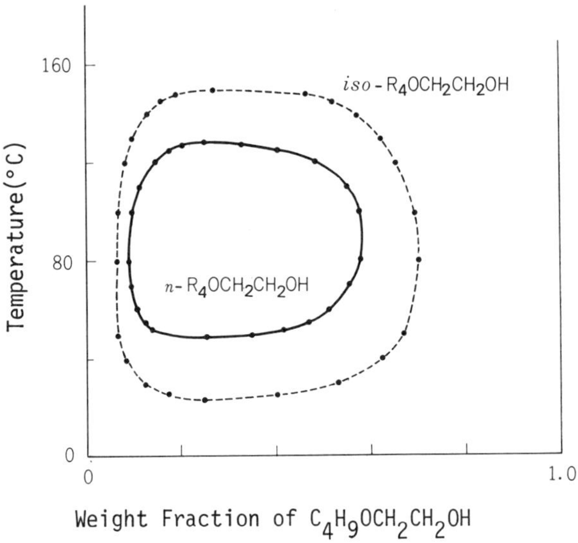

Fig. 5.10 Liquid-liquid solubility in the system H_2O-$C_4H_9OCH_2CH_2OH$.

eventually results. In some systems, such as nonionic surfactant-water or alkyl amine-water, the system splits into two phases at moderate temperatures, but mutual solubility increases with decreasing temperature, so that a single homogeneous phase is formed at the lower temperature. In such cases, the critical temperature is referred to as the lower consolute temperature. This phenomenon is the result of the strong hydration and dipole-dipole interactions and/or hydrate formation which are favored by low temperature. In some systems, such as water-nicotine, water-poly(ethylene glycol), and water-butyl monoglycol ether (a nonionic surfactant with a small molal volume), upper and lower critical points are observed over a moderate temperature range [11,12]. Examples are shown in Fig. 5.10 [13].

Sulfur solutions represent an unusual case. The sulfur exists as S_8 at temperatures below 159°C and forms regular solutions with hydrocarbons. The solubility increases with temperature, and a normal critical solution temperature is attained. However, above

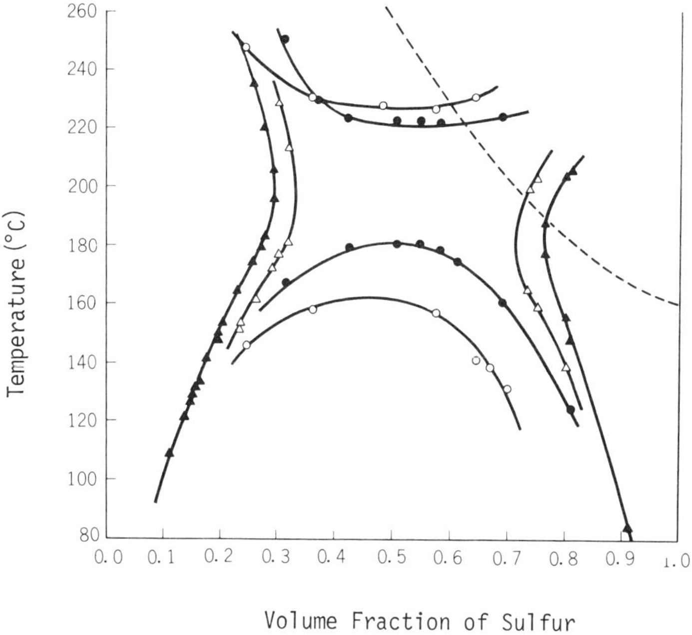

Fig. 5.11 Mutual solubility of C_6H_6-S (○); $C_6H_5CH_3$-S (●); m-$(CH_3)_2C_6H_4$-S (△); p-$(CH_3)_2C_6H_4$-S (▲). From Ref. 16. Reprinted with permission. Copyright by the American Chemical Society.

159°C, the amount of polymerized sulfur, S^{μ}, increases with increasing temperature, and thus the mutual solubility decreases, leading, finally, to separation into two phases. In other words, the solubility of sulfur decreases at both high and low temperatures. Examples of the solubility of sulfur in four organic solvents are shown in Fig. 5.11. The broken line in Fig. 5.11 indicates the amount of S^{μ} [14-16].

5.4 ESTIMATION OF THE SOLUBILITY OF SOLIDS IN LIQUIDS [17]

It is evident from Eq. (4.10) that the relative activity of the solid must be estimated in order to determine the solubility. In this case, the relative activity refers to the activity of the solid relative to that of the supercooled liquid at the temperature of interest. Employing the Clausius-Clapeyron equation, we obtain the enthalpy of vaporization and of sublimation,

$$\left(\frac{d \ln p_2^V}{dT}\right)_{sat} = \frac{\Delta H_2^V}{RT^2}; \qquad \left(\frac{d \ln p_2^S}{dT}\right)_{sat} = \frac{\Delta H_2^S}{RT^2}$$

Accordingly,

$$\frac{d \ln a_2^S}{dT} = \frac{d \ln p_2^S/p_2^\circ}{dT} = \frac{\Delta H_2^S - \Delta H_2^V}{RT^2} = \frac{\Delta H_2^F}{RT^2} \tag{5.23}$$

where ΔH_2^S is the heat of sublimation, ΔH_2^V the heat of vaporization, and ΔH_2^F the heat of fusion. The preceding equations may be integrated over the range from an arbitrary temperature T to the melting point T_f under the assumption that the heat of fusion is constant, yielding the familiar relationship

$$\ln a_2^S = -\frac{\Delta H_2^F}{R}\left(\frac{T_f - T}{TT_f}\right) \tag{5.24}$$

When the difference between the molal heat capacity of the liquid and the solid is known,

$$\Delta H_2^F(T) = \Delta H_2^F(T_f) - \Delta C_p(T_f - T) \tag{5.25}$$

where $\Delta C_p = C_p^\ell - C_p^S$.

Substituting Eq. (5.25) into Eq. (5.23), and integrating, we obtain

$$\ln a_2^S = -\frac{\Delta H_2^F}{R}\left(\frac{T_f - T}{TT_f}\right) + \frac{\Delta C_p}{R}\left(\frac{T_f - T}{T}\right) - \frac{\Delta C_p}{R} \ln \frac{T_f}{T} \tag{5.26}$$

It should be noted that ΔC_p varies with temperature and that considerably below the melting point, the value of a_2^S is subject to some uncertainty.

The following technique may be used to evaluate the relative activity of a solid substance for which the specific heat in both

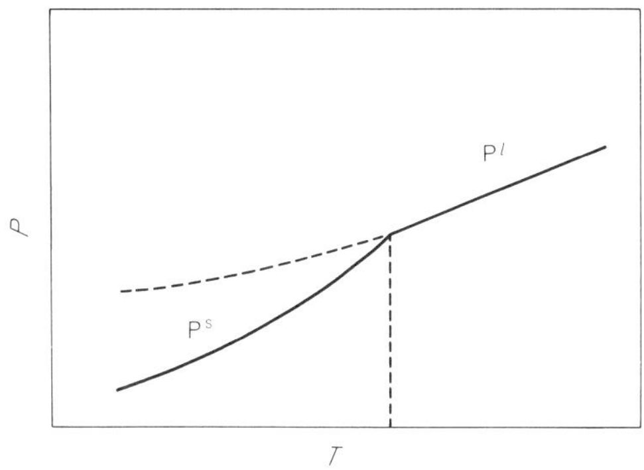

Fig. 5.12 Temperature variation of the vapor pressure of solid and supercooled liquid.

the solid and liquid states is not known. In a solution saturated with solid, the partial vapor pressure of the solute is equal to the vapor pressure of the solid. If we denote the vapor pressure of the supercooled liquid as p_2°, the relative activity of the solid under the conditions at which we wish to evaluate the solubility is $a_2 = p_2^S/p_2^\circ$. As indicated in Fig. 5.12, it is possible to extrapolate the curve corresponding to the liquid vapor pressure below the melting point to the desired temperature, and thus evaluate the ratio of the vapor pressures of the supercooled liquid and solid states. The solubility is then obtained by substituting a_2, V_2, δ_1, and δ_2 in the following equation:

$$\ln \frac{p_2^S}{p_2^\circ} \approx \ln a_2 \approx \ln x_2 + \frac{V_2 \phi_1^2 (\delta_1 - \delta_2)^2}{RT} \tag{5.27}$$

Table 5.4 illustrates selected relative activities and molal volumes for several solids at 25°C. The solubility parameter δ_2 of the solid must be determined from solubility data. However, once known, it can be used to estimate solubility in other solvents.

Similar compounds, such as anthracene and phenanthrene, have similar solubilities above their melting points. However, because the melting points differ significantly, their relative activities are quite different at temperatures below their melting points.

Table 5.4 Relative Activities and Molal Volumes (Liquid) for Solids at 25°C

Solids	a_2	V_2^ℓ (cm^3)
I_2	0.258	59
SnI_4	0.146	151
S_8	0.282	135
P_4	0.941	70
	0.865 (0°C)	

Phenanthrene is approximately 22 times as soluble as anthracene (see Table 11.1). The very low solubility of polytetrafluoroethane in any solvent is attributable to a very small relative activity, resulting from a high melting point as well as a large heat of fusion.

Thus, it is evident that the solubilities of gases, liquids, and solids are not different phenomena, but may be described by a single equation.

REFERENCES

1. J. M. Prausnitz, J. Phys. Chem., *66*, 640 (1962).
2. G. N. Lewis and M. Randall, *Thermodynamics*, 2d ed. revised by K. S. Pitzer and L. Brewer, McGraw-Hill, New York, 1961, pp. 187-188.
3. J. E. Jolley and J. H. Hildebrand, J. Am. Chem. Soc., *80*, 1050 (1958).
4. J. H. Hildebrand, J. M. Prausnitz, and R. L. Scott, *Regular and Related Solutions*, Van Nostrand Reinhold, New York, 1970, pp. 120-124, 218.
5. C. Black, G. G. Joris, and H. S. Taylor, J. Chem. Phys., *16*, 537 (1948).
6. K. Kinoshita, H. Ishikawa, and K. Shinoda, Bull. Chem. Soc. Japan, *31*, 1081 (1958).
7. C. McAuliffe, J. Phys. Chem., *70*, 1267 (1966).

8. J. H. Hildebrand, B. B. Fisher, and H. A. Benesi, J. Am. Chem. Soc., *72*, 7348 (1949).
9. K. Ishida, Bull. Chem. Soc. Japan, *31*, 143 (1958).
10. P. D. Schettler, Jr. and A. Patterson, Jr., J. Phys. Chem., *68*, 2865 (1964).
11. C. S. Hudson, Z. Phys. Chem., *47*, 113 (1903).
12. O. Flaschner, J. Chem. Soc., *95*, 668 (1909).
13. H. L. Cox and L. H. Cretcher, J. Am. Chem. Soc., *48*, 451 (1926).
14. H. R. Kruyt, Z. Phys. Chem., *65*, 486 (1909).
15. D. L. Hammick and W. E. Holt, J. Chem. Soc. (London), *1926*, 1955.
16. R. L. Scott, J. Phys. Chem., *69*, 261 (1965).
17. J. H. Hildebrand and R. L. Scott, *Solubility of Nonelectrolytes*, Reinhold, New York, 1950, pp. 270-299.

EXERCISES

5.1 m-Xylene and p-xylene form an ideal solution. p-Xylene has the higher melting point. At a temperature lower than the melting point of p-xylene, its crystals precipitate, and its solubility decreases with decreasing temperature. Calculate the solubility at 220.3 K, assuming that the heat of fusion is constant. Further, derive the solubility, considering the difference in molal heat capacities between supercooled liquid and solid ΔC_p and compare with previous result. For p-xylene, T_f = 286.39 K, ΔH_f = 4090 cal/mol, ΔC_p = 3R = 5.96 cal/deg, and for m-xylene T_f = 219.5 K.

Answer: From Eq. (5.24)

$$\ln a_2^S = \ln x_2 = \frac{\Delta H_2^F}{R}\left(\frac{T - T_f}{TT_f}\right) = \frac{4090(-66.1)}{1.987(220.3)(286.4)} = -2.16$$

$$x_2 = 0.1161$$

From Eq. (5.26)

$$\ln a_2^S = \ln x_2 = \frac{(\Delta H_2^F - T_f\Delta C_p)(T - T_f)}{RTT_f} - \frac{\Delta C_p}{R} \ln \frac{T_f}{T}$$

$$= \frac{(4090 - 286.4 \times 5.96)(-66.1)}{1.987(220.3)(286.4)} - 3 \ln \frac{286.4}{220.3}$$

$$= -2.045$$

$x_2 = 0.1295$

5.2 Calculate f/p for N_2 at 50 atm and 273 K, using the Berthelot equation. For N_2, $T_c = 126.1$ K and $P_c = 33.5$ atm.

Answer: The fugacity of a gas which follows Berthelot's equation is given by Eq. (5.8). Substituting $T_c = 126.1$ K, $P_c = 33.5$ atm for N_2, at P = 50, T = 273 K

$$\ln \frac{f}{P} = \frac{9T_cP}{128TP_c}\left(1 - 6\frac{T_c^2}{T_2}\right)$$

$$= \frac{9(126.1)(50)}{128(273)(33.5)}\left(1 - 6\frac{(126.1)^2}{(273)^2}\right)$$

$$= -1.34 \times 10^{-2}$$

$$\frac{f}{P} = 0.9867$$

5.3 Calculate the solubility of solid SnI_4 in liquid $SiCl_4$ at 298 K. The experimental value of $V_2(\delta_1 - \delta_2)^2/RT = 3.8$. The melting point and heat of fusion of SnI_4 is 417.7 K and 4.48 K cal/mol, respectively. Assume $\Delta C_p = 0$.

Answer: In a saturated solution the following equations hold:

$$\mu_2(\text{solute}) = \mu_2^\circ(1) + RT \ln x_2 + V_2\phi_1^2(\delta_1 - \delta_2)^2$$

$$\mu_2(\text{solute}) = \mu_2^\circ(s) + RT \ln x_2 + V_2\phi_1^2(\delta_1 - \delta_2)^2 + \Delta H_f - T\frac{\Delta H_f}{417.7}$$

$$\mu_2(\text{solute}) = \mu_2^\circ(s)$$

Assuming $\phi_1 \approx 1$ we obtain

$$\ln x_2 = -3.8 - 7.56 + 5.39$$

$$x_2 = 0.0026$$

As x_2 is very small, the calculation carried out with the assumption $\phi_1 \approx 1$ is correct.

5.4 The chemical potential of the second component in a solute state $\bar{G}_2$ is equal to that in equilibrium with the solution G_2° (e.g., solid or gas). Derive the following equation, under the condition p = constant.

$$\left[\frac{\partial(\bar{G}_2 - G_2^\circ)}{\partial T}\right]_x = -\left[\frac{\partial(\bar{G}_2 - G_2^\circ)}{\partial \ln x_2}\right]_T \left(\frac{\partial \ln x_2}{\partial T}\right)_{\bar{G}_2 - G_2^\circ}$$

Answer: Total differentiation of $G_2 - G_2^\circ$ yields (p being constant)

$$d(\bar{G}_2 - G_2^\circ) = \left[\frac{\partial(\bar{G}_2 - G_2^\circ)}{\partial T}\right]_x dT + \left[\frac{\partial(\bar{G}_2 - G_2^\circ)}{\partial \ln x_2}\right]_T d \ln x_2$$

Partial differentiation with respect to T under the equilibrium condition, i.e., $\bar{G}_2 - G_2^\circ$ = constant = 0, yields

$$\left[\frac{\partial(\bar{G}_2 - G_2^\circ)}{\partial T}\right]_{\bar{G}_2 - G_2^\circ} = 0 = \left[\frac{\partial(\bar{G}_2 - G_2^\circ)}{\partial T}\right]_x + \left[\frac{\partial(\bar{G}_2 - G_2^\circ)}{\partial \ln x_2}\right]_T \left(\frac{\partial \ln x_2}{\partial T}\right)_{\bar{G}_2 - G_2^\circ}$$

PROBLEMS

5.1 From the basic solubility equation (4.10)

$$RT \ln a_2 = RT \ln x_2 + V_2\phi_1^{\,2}(\delta_1 - \delta_2)^2$$

calculate the critical temperature and composition in the liquid-liquid mixture.

5.2 The change of the solubility of I_2 in $CCl_2F \cdot CClF_2$ with temperature is shown in the table below. Calculate the entropy of solution of I_2 at 25°C, making the approximation that the change of activity is proportional to the change in concentration.

Temperature (°C)	*Solubility (mol%) ($100\ x_2$)*
4.86	0.1007
14.95	0.1601
20.00	0.1968
25.09	0.2474
29.97	0.2985

chapter six

ENTHALPY AND ENTROPY OF SOLUTION

6.1 ENTHALPY AND ENTROPY OF SOLUTION OF SOLIDS

It was shown in Chap. 5 that the solubility of gases, liquids, or solids in liquids could be equally well described by Eq. (4.10). Other systems which can be similarly treated are liquids or solids in gas or vacuum. The vapor pressure of the liquids or solids may be regarded as the solubility in vacuum, which is a fluid whose intermolecular force is zero.

Extending further the principle of the basic solubility equation, the following development is possible. Substances with small vapor pressures generally have large heats of vaporization, and are thus usually large molecules. Substances with noticeable smells have a more or less significant vapor pressure, and it may be expected that they could be purified by molecular or vacuum distillation, or by vacuum sublimation. On the other hand, polymers must be odorless, and it may easily be guessed that they cannot be purified by distillation. Any odors in a polymer must be due to impurities or the products of decomposition. Generally, solid inorganic substances, including stones and rocks, are odorless, thus indicating that they are not the molecular solids indicated by their chemical formulas, but rather polymeric. Organic compounds generally are maintained in the liquid or solid state by the dispersion and dipole moment forces between the molecules, while inorganic compounds form macromolecules through chemical bonds, e.g., covalent, ionic, or coordination. This difference may be explained by noting that the carbon atoms, which form the framework of organic compounds, are equipped with four valence bonds and four bond orbitals. On the other hand, in many inorganic substances there are more bond orbitals than valence bonds. For example, aluminum has four bond orbitals, but only three valence bonds; hence, compounds containing aluminum show polymer characteristics owing to the formation of coordinate bonds.

The heat of vaporization of substances with an appreciable vapor pressure can be obtained by means of the Clausius-Clapeyron equation, using the vapor pressure change with temperature. By considering the vapor pressure as the solubility in vacuum, the heat of vaporization may be regarded as the heat of solution into vacuum. Thus, in the same way, we may use the Clausius-Clapeyron equation, substituting solubility for the vapor pressure, to calculate the enthalpy of solution of solids in liquids.

The chemical potential μ_2 of the solute in a saturated solution is equal to the Gibbs energy $G_2{}^s$ of the pure solid phase separated,

$$\bar{G}_2 - G_2{}^s = 0 \qquad (\mu_2 = \bar{G}_2) \tag{6.1}$$

Accordingly,

$$\bar{S}_2 - S_2{}^s = \frac{\bar{H}_2 - H_2{}^s}{T} = -\left[\frac{\partial(\bar{G}_2 - G_2{}^s)}{\partial T}\right]_{x_2} \tag{6.2}$$

By partial differentiation, as was done with Eq. (5.10),

$$-\left(\frac{\partial(\bar{G}_2 - G_2{}^s)}{\partial T}\right)_{x_2} = \left[\frac{\partial(\bar{G}_2 - G_2{}^s)}{\partial \ln x_2}\right]_T \left(\frac{\partial \ln x_2}{\partial T}\right)_{\bar{G}_2 - G_2{}^s} \tag{6.3}$$

When $\bar{G}_2 - G_2{}^s$ is separated into terms corresponding to free energy of melting, $G_2^\circ - G_2{}^s$, and free energy of solution, $\bar{G}_2 - G_2^\circ$, the differential of the first term is zero, since the free energy of melting is not a function in concentration of solute. By the

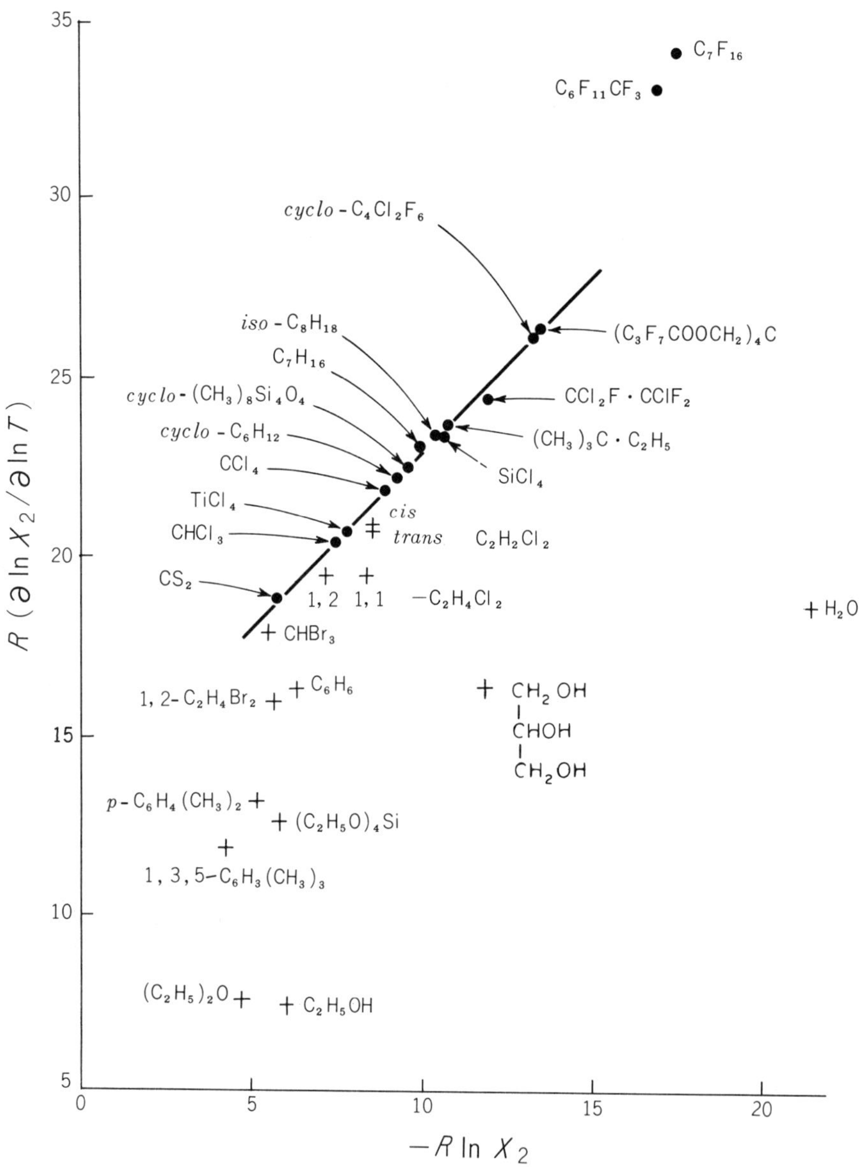

Fig. 6.1 Relation between solubility and temperature coefficient of solubility for iodine in various solvents. Regular solutions are designated by •.

definition of relative activity, $\bar{G}_2 - G_2^\circ = RT \ln a_2$, and therefore

$$\left(\frac{\partial(\bar{G}_2 - G_2{}^s)}{\partial \ln x_2}\right)_T = RT\left(\frac{\partial \ln a_2}{\partial \ln x_2}\right)_T \tag{6.4}$$

Substituting Eqs. (6.3) and (6.4) into Eq. (6.2)

$$\bar{S}_2 - S_2{}^s = \frac{\bar{H}_2 - H_2{}^s}{T} = RT\left(\frac{\partial \ln a_2}{\partial \ln x_2}\right)_T \left(\frac{\partial \ln x_2}{\partial T}\right)_{\bar{G}_2 - G_2{}^s} \tag{6.5}$$

Therefore, the entropy and heat enthalpy of solution may be calculated from the temperature dependence of solubility and the change of relative activity with concentration [1]. The term involving the change in relative activity with concentration is exactly similar to the change of fugacity with vapor pressure, and has a value close to unity at low concentration.

Figure 6.1 illustrates the relation between the solubility of iodine and the temperature coefficient of solubility in a number of solvents.

There is a charge-transfer interaction between iodine (electron acceptor) and aromatics, olefins, ethers, and ketones (electron donors). As a result, such systems differ from regular solutions, and do show the nonviolet color. These solutions are indicated by +. The filled circles in Fig. 6.1 indicate solvents which form regular solutions with iodine, and such solutions are violet in color, as is iodine vapor. It is evident from Fig. 6.1 that the change of solubility with temperature (entropy of solution) increases by the same amount as the solubility decreases, i.e., the entropy of mixing $-R \ln x_2$ increases. The solubility of iodine in nonviolet solvents with charge-transfer interaction is much greater than would be estimated from the solubility parameter, and the entropy of solution shows a negative deviation from the violet solutions (solid line).

6.2 CHANGE IN ACTIVITY WITH CONCENTRATION

Insufficient attention has been paid to the change in activity with concentration, in spite of the fact that it is the activity, and not the concentration, which is used in thermodynamic formulations. The theory has been developed by taking the standard of activity in such a way that the activity in infinitely dilute solution is equal to the concentration, since the activity is proportional to the concentration in dilute solutions. However, we must be careful to avoid the mistake of assuming that the activity changes proportionately with the concentration in dilute solutions without a complete understanding of the phenomenon of solution, as well as the different mechanism involved in surfactant solutions and in solubilization. With respect to this latter phenomenon, the reader is referred to Chap. 9, in which surfactant solutions are discussed. In ordinary solutions, where such phenomena as polymolecular association are absent, the change in activity with concentration may be obtained as follows, if the solubility or partial pressure is known [2].

Using the basic solubility equation

$$\ln a_2 = \ln x_2 + \frac{V_2\phi_1^2(\delta_1 - \delta_2)^2}{RT} \tag{4.10}$$

and differentiating at constant P and T

$$\left(\frac{\partial \ln a_2}{\partial \ln x_2}\right)_{P,T} = 1 - \frac{2\phi_2}{x_1}\frac{V_2\phi_1^2(\delta_1 - \delta_2)^2}{RT} \tag{6.6}$$

Substituting into Eq. (4.10), we obtain

$$\left(\frac{\partial \ln a_2}{\partial \ln x_2}\right)_{P,T} = 1 - 2\,\frac{\phi_2\phi_1^2}{x_1\phi_{1\,\mathrm{sat}}^2}\ln\frac{a_{2\,\mathrm{sat}}}{x_{2\,\mathrm{sat}}} \tag{6.7}$$

Equations (6.6) and (6.7) are convenient, since they are applicable at any concentration in a simply dispersed (i.e., nonassociated solute) solution. In the case of solutions which behave like ideal solutions, the second term becomes nearly zero, and thus the left-hand side becomes equal to unity. In the case of low solubility, the left-hand side approaches unity, since ϕ_2 becomes small.

Equations (6.6) and (6.7) always have positive values smaller than one for actual solutions, and become equal to zero only for the critical composition at the critical temperature, as was shown to be the case for the mutual solution of liquids. In fact, at the critical temperature of the system, where liquids with equal molal volumes mutually dissolve,

$$V_2(\delta_1 - \delta_2)^2 = 2RT, \qquad \phi_1 = 0.5$$

Substituting into Eq. (4.10), we obtain

$$\ln \frac{a_2}{x_2} = 0.5$$

and further substitution into Eq. (6.6) leads to

$$\frac{\partial \ln a_2}{\partial \ln x_2} = 0$$

When the heat of solution or the partial vapor pressure is known, the quantity $B = V_2(\delta_1 - \delta_2)^2$ can be obtained and we obtain

$$\left(\frac{\partial \ln a_2}{\partial \ln x_2}\right)_{P,T} = 1 - 2\,\frac{\phi_2}{x_1}\,\frac{B\phi_1^2}{RT} \qquad (6.6')$$

Thus, from Eq. (6.6'), the activity change with concentration at any composition may be calculated with a fair degree of precision.

The change in activity with concentration of iodine solutions is shown in Table 6.1. As the table indicates, the change in activity with concentration is always smaller than one, and is mainly in the range 0.85 to 0.99. Equation (6.7) is extremely useful in a system of low solubility, provided that the solubility and the relative activity are known, since the precise measurement of vapor pressure or heat of mixing is difficult in the dilute solution of a solid [2]. The entropy of solution of iodine in various

Table 6.1 Solubility, Temperature Dependence of Solubility, Change of Activity with Concentration at Saturation and Entropy of Solution of Iodine in Various Solvents at 25°C
(a_2 = 0.258)

Solvents	$100x_2$	$-R \ln x_2$	$R\left(\frac{\partial \ln x_2}{\partial \ln T}\right)$	$\frac{\partial \ln a_2}{\partial \ln x_2}$	$\bar{S}_2 - S_2^s$
C_6H_6	4.80	6.04	16.3	0.886	14.4
$CHBr_3$	6.25	5.51	17.8	0.870	15.5
CS_2	5.58	5.73	18.8	0.824	15.5
$CHCl_3$	2.34	7.50	20.3	0.916	18.6
1,2-$C_2H_4Br_2$	7.82	5.07	15.9	0.863	13.7
1,2-$C_2H_4Cl_2$	2.20	7.60	19.5	0.917	17.9
CH_2Cl_2	1.52	8.32	20.4	0.920	18.8
n-$C_{16}H_{34}$	1.45	8.42	20.6	0.983	20.3
CCl_4	1.15	8.88	21.9	0.956	20.9
n-C_7H_{16}	0.68	9.92	23.1	0.98	22.6
cyclo-$(CH_3)_8Si_4O_4$	0.81	9.57	22.5	0.990	22.3
$TiCl_4$	2.15	7.63	20.7	0.942	19.5
cyclo-C_6H_{12}	0.92	9.32	22.2	0.967	21.5
$SiCl_4$	0.499	10.5	23.3	0.98	22.8
$CCl_2F{\cdot}CClF_2$	0.245	11.9	24.4	0.99	24.1
cyclo-$C_4Cl_2F_6$	0.124	13.3	26.2	0.995	26.1
$(C_3F_7COOCH_2)_4C$	0.117	13.4	26.3	0.999	26.3
cyclo-$C_6F_{11}{\cdot}CF_3$	0.021	16.8	31.7	0.997	31.6
n-C_7F_{16}	0.0184	17.1	34.4	0.998	34.3

regular solvents at constant pressure obtained from Table 6.1, are plotted versus the entropy of mixing $-R \ln x_2$ in Fig. 6.2 (closed circles). The slope of the dotted line is no longer 1 but 1.1 to 1.4. However, values of fluorocarbons lie on this curve and the intersection of this curve with the ordinate, 9 eu, is close to the entropy of fusion of iodine.

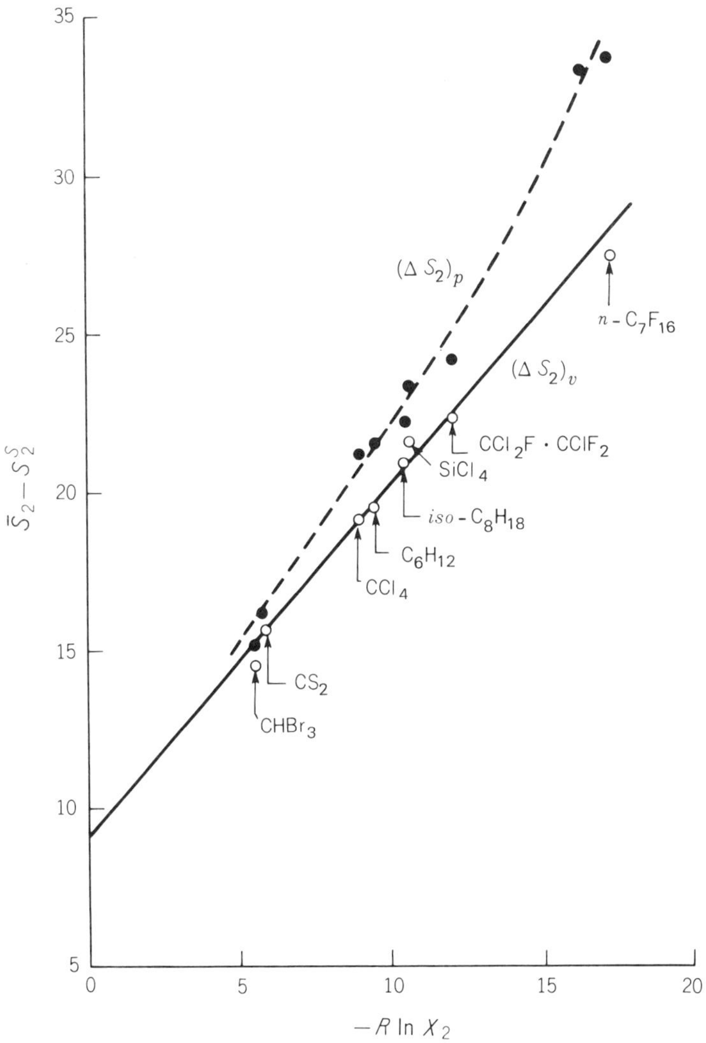

Fig. 6.2 Entropy of solution of iodine at constant pressure (•) and at constant volume (○). From Ref. 4. Reprinted with permission. Copyright by the American Chemical Society.

6.3 ENTROPY OF SOLUTION AT CONSTANT PRESSURE AND VOLUME

The most readily determined values, such as enthalpy of solution, entropy of solution, and heat capacity are those measured at constant pressure. On the other hand, the theoretical values are usually derived under the condition of constant volume. Therefore, it is necessary to know the difference in the two quantities thermodynamically, and thus obtain the value at constant volume from the experimental value at constant pressure. For example, in the case of specific heat, one may calculate the specific heat at constant volume from the experimentally observed value at constant pressure by means of the known thermodynamic relation

$$C_V = C_P - T\left(\frac{\partial P}{\partial T}\right)_V\left(\frac{\partial V}{\partial T}\right)_P$$

where the quantities $(\partial P/\partial T)_V$ and $(\partial V/\partial T)_P$ are both observable.

The change of volume accompanying a process at constant pressure is $V_m - V°$, and the change of entropy of system with volume is $(\partial S/\partial V)_T = (\partial P/\partial T)_V$. Therefore,

$$(\Delta S_m)_P - (\Delta S_m)_V = \left(\frac{\partial S}{\partial V}\right)_T \quad \Delta V = \left(\frac{\partial P}{\partial T}\right)_V(V_m - V°) \tag{6.8}$$

where V_m is the volume of solution mixed at constant pressure and V is the volume before mixing.

The partial molal entropy of solution can be obtained by partial differentiation with respect to component 2:

$$(\Delta\bar{S}_2)_P - (\Delta\bar{S}_2)_V = \left(\frac{\partial P}{\partial T}\right)_V(\bar{V}_2 - V_2) \tag{6.9}$$

In dilute solutions $(\partial P/\partial T)_V$ for the solution is approximately equal to that of the solvent. On the other hand, while $V_P^M - V° \approx 0$, the change in partial molal volume of the solute is not zero. Table 6.2 shows the partial molal volumes $\bar{V}_2$ of iodine in various solvents in dilute solution [3].

As the molal volume of supercooled liquid iodine at 25°C is 59 cm^3, $\bar{V}_2 - V_2$ can be calculated. When this quantity is multiplied

Table 6.2 Partial Molal Volumes of Iodine in Various Solvents at 25°C

Solvents	$\bar{V}_2$ (cm^3)	Solvents	$\bar{V}_2$ (cm^3)
n-C_7F_{16}	100	cyclo-$(CH_3)_8Si_4O_4$	66.6
cyclo-$C_4Cl_2F_6$	81.2	cyclo-C_6H_{12}	68.2
CCl_2F $CClF_2$	67.7	CCl_4	66.7
$SiCl_4$	67.1	$CHCl_3$	65.6
iso-C_8H_{15}	66.7	CS_2	60.7
n-C_7H_{16}	66.3	$I_2(\ell)$	59

by $(\partial P/\partial T)_V$ for the solvent, the entropy of solution at constant volume is obtained from that at constant pressure by means of Eq. (6.9), as is shown as the open circles in Fig. 6.2.

The slope of $R(\partial \ln x_2/\partial \ln T)$ vs. $-R \ln x_2$ was seen in Fig. 6.1 to be almost equal to unity, with the exception of the fluorocarbon solvents. Correction of the change in activity with concentration yields the slope of $(\Delta\bar{S}_2)_p$ vs. $-R \ln x_2$ somewhat steeper than unity, which is apparently difficult to explain; however, the data for the fluorocarbon solvents were included on the dotted curve in Fig. 6.2. Further correction for change in volume yields the slope of $(\Delta\bar{S}_2)_V$ vs. $-R \ln x_2$ close to unity, i.e., 1.1. Thus, it is evident that the entropy of solution increases to the same extent that the solubility decreases, i.e., the entropy of dilution increases. When the slope of entropy at constant volume is extrapolated to the concentration $x_2 = 1$, the difference in entropy of solution at constant volume and the entropy of mixing is seen to be approximately 9 eu. This is nearly equal to the entropy of fusion of iodine. It thus appears that the entropy of solution of solids in liquids in regular solutions, both at constant volume and constand pressure, can be approximated as follows [3,4]:

$$(\Delta\bar{S}_2)_V \approx -R \ln x_2 + \Delta S_{2f} \qquad \text{(solid} \rightarrow \text{solute)} \qquad (6.10)$$

$$(\Delta\bar{S}_2)_P \approx -R \ln x_2 + \Delta S_{2f} + \left(\frac{\partial P}{\partial T}\right)_V (\bar{V}_2 - V_2) \qquad \text{(solid} \rightarrow \text{solute)} \qquad (6.11)$$

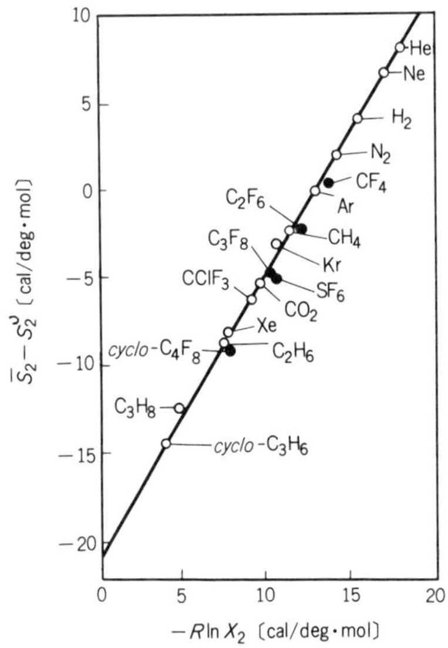

Fig. 6.3 Entropy of solution of various gases in cyclo-hexane (25°C, 1 atm). From Hildebrand/Prausnitz/Scott: *Regular and Related Solutions*. Copyright 1970 by Litton Educational Publishing Inc. Published by Van Nostrand Reinhold Company.

The enthalpy of solution of solids in liquids in regular solution at constant volume can be approximated:

$$(\Delta\bar{H}_2)_V \approx -RT \ln \frac{x_2}{a_2} + \Delta H_2^F \qquad \text{(solid} \rightarrow \text{solute)} \qquad (6.12)$$

Similar relations are found when stannic iodide is dissolved in various solvents [5]. In the case of sparingly soluble liquids in liquids, there is no fusion term:

$$(\Delta\bar{S}_2)_V \approx -R \ln x_2 \qquad \text{(pure liquid} \rightarrow \text{solute)} \qquad (6.13)$$

$$(\Delta\bar{H}_2)_V \approx -RT \ln x_2 \qquad \text{(pure liquid} \rightarrow \text{solute)} \qquad (6.14)$$

In the case of gases in liquids, there is a condensation term. Some results are shown in Fig. 6.3.

$$(\Delta\bar{S}_2)_V \approx -R \ln x_2 - \Delta S_2^V \qquad \text{(gas} \rightarrow \text{solute)} \qquad (6.15)$$

$$(\Delta\bar{H}_2)_V \approx -RT \ln \frac{x_2}{a_2} - \Delta H_2^V \qquad \text{(gas} \rightarrow \text{solute)} \tag{6.16}$$

Thus it may be seen that a measurement of the solubility in regular solution at an arbitrary temperature permits the calculation of an approximate value of the temperature coefficient of solubility and of the entropy of solution.

6.4 ENTROPY OF SOLUTION IN SYSTEMS WITH MARKEDLY DIFFERENT MOLAL VOLUMES

In Sec. 2.2, the entropy of mixing was calculated under the assumption that the free volume of each component in solution, as well as in the pure liquid, were equal:

$$\Delta\bar{S}_2 = -R \ln x_2 \tag{2.17}$$

On the further assumption that the free volumes of the components are proportional to their molal volumes:

$$\Delta\bar{S}_2 = -R\left[\ln \phi_2 + \phi_1\left(1 - \frac{V_2}{V_1}\right)\right] \tag{2.18}$$

Let us now consider what sort of model would be required to describe the behavior of actual solutions. In ordinary solutions there is not a great difference in the molal volumes of the components. A typical small volume would be 60 cm^3 at 25°C for CS_2, and an example of a fairly large volume would be 215 cm^3 at 25°C for n-C_7F_{16}, among those compounds which are liquid at normal temperatures. Such molecules are in constant thermal motion, and are undergoing rotation as well. Consequently, they may be approximated as approximately spherical rigid bodies.

The thermodynamically defined free volume from the entropy of vaporization,

$$\Delta S^V = R \ln \frac{V_g}{V_f} \tag{1.12}$$

is almost the same as that defined from the free space available to

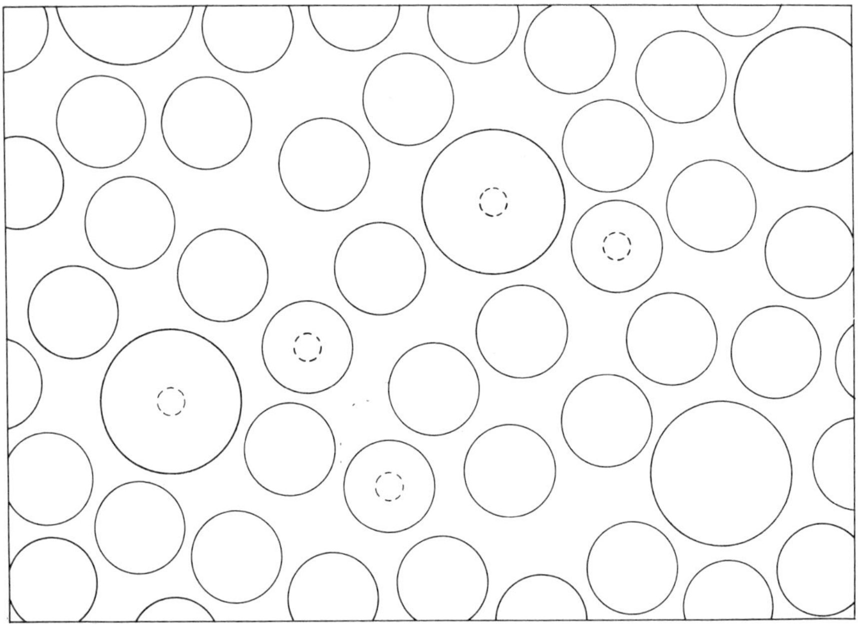

Fig. 6.4 Schematic representative of the fact that the free volume, i.e., the free space, available to the center of mass of the molecules is the same regardless of the size of the molecules.

the center of gravity of the molecules which make up the liquid as discussed in Chap. 1. Accordingly, when a solution is formed by mixing of two liquids, the space available for free motion of the center of gravity of the large molecules and of the small molecules are almost identical, as shown schematically in Fig. 6.4. If there are not too many molecules of solute, the free volumes of solvent and solute may be taken as equal to the free volume of the solvent.

Substituting $V_{1f} = \bar{V}_{1f} = \bar{V}_{2f}$ in Eq. (2.16)

$$\Delta S^M = -R\left(n_1 \ln \frac{n_1}{n_1 + n_2} + n_2 \ln \frac{n_2}{n_1 + n_2} \frac{V_{2f}}{V_{1f}}\right) \tag{6.17}$$

where V_{1f} is the free volume of component 1, and $\bar{V}_{1f}$ and $\bar{V}_{2f}$ are the partial molal free volumes of the components of the solution.

The partial molal entropy of mixing may be obtained by differentiation of Eq. (6.17) with respect to n_2

$$\Delta\bar{S}_2 = \left(\frac{\partial \Delta S_m}{\partial n_2}\right)_{n_1} = -R\left(\ln x_2 + \ln \frac{V_{2f}}{V_{1f}}\right) \tag{6.18}$$

Since the free volumes of the components are not greatly different, we may ignore the second term in the parentheses, and we then obtain the same relation which applies to ideal solutions. If the precise values of V_{1f} and V_{2f} are introduced in Eq. (6.18), a more precise result is obtained. However, precise values of the free volume are not generally available.

In Fig. 6.1, the molal volumes of solvents are markedly different: $(C_3F_7COOCH_2)_4C$ = 540 cm^3; cyclo-$C_4Cl_2F_6$ = 140; $SiCl_4$ = 115; iso-C_8H_{18} = 166; cyclo-$(CH_3)_8Si_4O_4$ = 312; cyclo-C_6H_{12} = 104; CCl_4 = 97. As the molal volume of iodine, 59 cm^3, is small, the entropy of mixing calculated through the use of Eq. (2.18) yields a larger entropy of mixing than that calculated by the use of Eq. (2.17), i.e., 0.15 eu for CCl_4, 1.6 eu for $(CH_3)_8Si_4O_4$, and 2.6 eu for $(C_3F_7COOCH_2)_4C$. However, the experimental results do not show the deviations expected from the different solvent molal volumes [6]. We must thus conclude that the model employed, in which the free volume is assumed to be proportional to the molal volume, is not correct. Equation (2.18) is identical with that for the entropy of mixing in the Flory-Huggins theory of polymer solution. It would be expected that this equation should give values which are too high for the entropy of mixing in ordinary solutions. In order to make this important conclusion clear, another experimental result will be cited.

The liquid-liquid solubility curves in the systems composed of two markedly different molal volumes are shown in Fig. 6.5. The asymmetry is so extreme in these cases as to afford an additional basis for deciding between ideal entropy and Flory-Huggins entropy of mixing [7]. The theoretical composition derived, assuming the ideal entropy of mixing, is

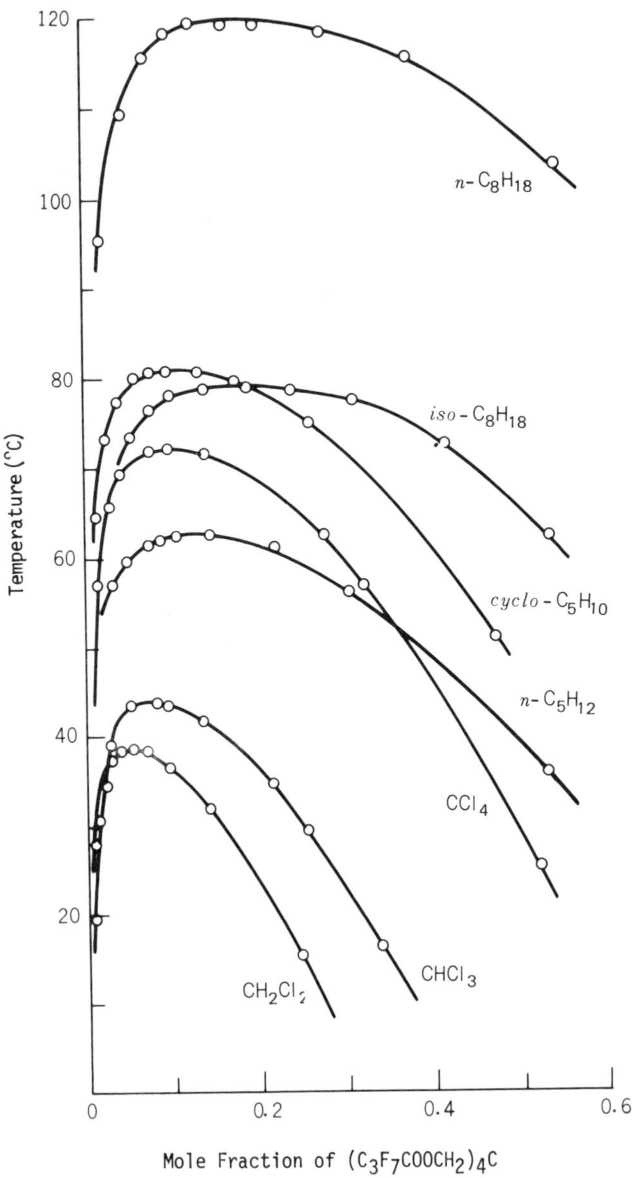

Fig. 6.5 Liquid-liquid solubility curves for molecules of very different sizes. From Ref. 7(b). Reprinted with permission. Copyright by the American Chemical Society.

Table 6.3 Comparison of Experimental and Calculated Critical Compositions in Liquid Mixtures of Large Molar Volume, $(C_3F_7COOCH_2)_4C$, and of Small Molar Volume, V_1

Liquids	T°C	V_1 (cm^3)	V_2 (cm^3)	x_2 (expt)	$\frac{V_2 - (V_1^2 - V_1V_2 + V_2^2)^{1/2}}{V_2 - V_1}$	$\frac{V_1^{3/2}}{V_2^{3/2} + V_1^{3/2}}$
CH_2CL_2	38.5	65.7	549.5	0.057	0.061	0.040
$CHCl_3$	43.5	82.6	553	0.073	0.077	0.055
n-C_5H_{12}	62.1	123.8	564	0.13	0.115	0.093
CCl_4	72.1	103	571	0.09	0.094	0.079
cyclo-C_5H_{10}	80.8	101	575	0.10	0.092	0.069
iso-C_8H_{18}	78.8	178	573	0.17	0.164	0.147
n-C_8H_{18}	119.4	185	603	0.16	0.162	0.145

$$x_2 = \frac{V_2 - (V_1^2 - V_1V_2 + V_2^2)^{1/2}}{V_2 - V_1} \qquad (5.16)$$

and that derived assuming Flory-Huggins entropy of mixing is

$$\frac{x_2}{x_1} = \left(\frac{V_1}{V_2}\right)^{3/2} \qquad (5.22)$$

As can be seen from Table 6.3, Eq. (5.16), i.e., $-R \ln x_2$, for the entropy of mixing gives the more satisfactory result [7].

6.5 VOLUME CHANGES ON MIXING

Of the various thermodynamic functions for the mixing process, the volume change on mixing seems most easily influenced by the specific shapes of the constituent molecules, and therefore cannot be predicted accurately by a simple theory. It is important to know the volume change, especially of the partial molal volume, not only for its intrinsic interest, but for the information this gives us about the state of solution of solute. In addition, this information is needed in order to calculate the entropy of solution at constant volume from those at constant pressure.

It is known that the value of the partial pressure for many solutions is higher than that for ideal solutions, and that the volume increases when the solution process is endothermic.

The relation between the volume change and the excess free energy of mixing is given by the following equation [8]:

$$V^E \approx N\beta E_V = n\beta F^E \qquad (6.19)$$

while the partial molal quantities are

$$V_2^E = \bar{V}_2 - V_2 \approx n\beta\Delta\bar{E}_{2V} \approx n\beta F_2^E \qquad (6.20)$$

where β is the compressibility, and

$$n = -\left(\frac{\partial E}{\partial V}\right)_T \frac{V}{E_V}$$

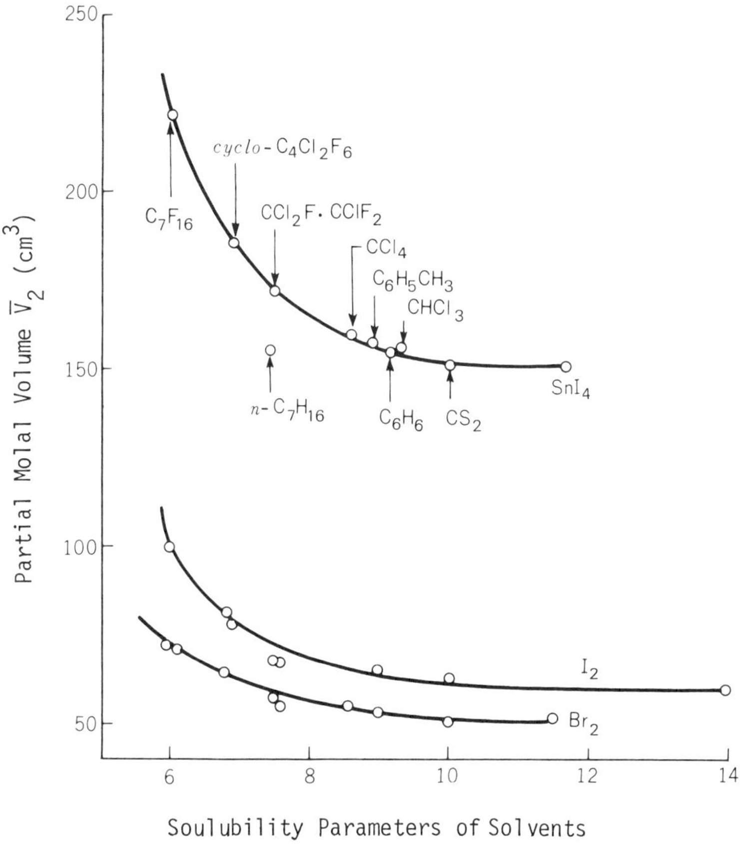

Fig. 6.6 Partial molal volumes versus solubility parameters for various solvents. From Ref. 9. Reprinted with permission. Copyright by the American Chemical Society.

From Eqs. (6.19) and (6.20) we see that the volume increases in an endothermic reaction, i.e., $\Delta\bar{E} > 0$, and vice versa.

In systems in which there is complete mutual solubility of each component, the excess free energy is not large, and hence the change in volume and partial molal volume are both small. However, in the case where the heat of mixing is large, i.e., where mutual solubility is limited, the change in partial molal volume will be fairly large.

In Fig. 6.6 is shown the relation between the partial molal volumes of I_2, Br_2, and SnI_4 in infinitely dilute solution and the

Table 6.4 Partial Molal Volumes, V_2, and Excess Volumes, $V_2^E = V_2 - V_2^\circ$, in n-C_7F_{16}

Solutes	V_2 (cm^3)	V_2°	$\bar{V}_2^E$ (expt)	$\bar{V}_2^E$ (calc)
I_2	100 [10]	59	41	60
Br_2	73. [11]	51.5	21.5	24.7

solubility parameter of solvents. From this figure, it is evident that the partial molal volume increases substantially when the solvent changes from good to poor solvents [3,9].

Table 6.4 compares the theoretical with the observed value for the case of large increase of partial molal volume. In the table, the theoretical values are obtained by using the activity coefficient, γ_2, derived from the solubilities of I_2 or Br_2, employing $n\beta = 13.9 \times 10^{-3}$ cm^3/cal, with $n = 1.49$, $\beta = 2.34 \times 10^{-4}$ atm for C_7F_{16}. We thus see that Eq. (6.15) does describe the large changes in partial molal volume in a semiquantitative way.

Later, Hildebrand and Dymond [12] proposed another approach. It is assumed that introducing solute molecules into the solvent causes the formation of microscopic regions of weakness and expansion, until enough energy has been absorbed by the system to restore the local pressure to the bulk internal pressure. Accordingly, we write

$$\left(\frac{\partial E}{\partial V}\right)_T (\bar{V}_2 - V_2) = \bar{E}_2 - E_2^\circ \tag{6.21}$$

or

$$\left(\frac{\partial E}{\partial V}\right)_T (\bar{V}_2 - V_2) = V_2\phi_1^{\,2}(\delta_1 - \delta_2)^2 \tag{6.22}$$

These equations give better agreement with observed values for liquid-liquid mixtures [12].

In the case of mutually soluble components, the partial molal volumes can usually be determined by measurement of density versus concentration. However, sparingly soluble components (or for solutes whose partial molal volumes change irregularly over a narrow concentration range), a direct dilatometric measurement [3], or measurements with a magnetic density balance [13] are preferable.

REFERENCES

1. J. H. Hildebrand, J. Chem. Phys., *20*, 190 (1952).
2. K. Shinoda and J. H. Hildebrand, J. Phys. Chem., *61*, 789 (1957); *69*, 607 (1965).
3. K. Shinoda and J. H. Hildebrand, J. Phys. Chem., *62*, 295 (1958).
4. J. H. Hildebrand, J. Phys. Chem., *64*, 370 (1960).
5. E. B. Smith and J. Walkley, Trans. Faraday Soc., *56*, 1276 (1960).
6. K. Shinoda and J. H. Hildebrand, J. Phys. Chem., *61*, 789 (1957); *62*, 292 (1958).
7. K. Shinoda and J. H. Hildebrand, J. Phys. Chem., *63*, 48 (1958); *65*, 1885 (1961).
8. J. H. Hildebrand and R. L. Scott, *Regular Solutions*, Prentice-Hall, Englewood Cliffs, N. J., 1962, pp. 106-109.
9. E. B. Smith, J. Walkley, and J. H. Hildebrand, J. Phys. Chem., *63*, 703 (1959).
10. D. N. Glew and J. H. Hildebrand, J. Phys. Chem., *60*, 616 (1956).
11. L. W. Reeves and J. H. Hildebrand, J. Phys. Chem., *60*, 949 (1956).
12. J. H. Hildebrand and J. Dymond, J. Phys. Chem., *46*, 624 (1967).
13. L. Benjamin, J. Phys. Chem., *70*, 3790 (1966).

EXERCISES

6.1 Examine the relation among vapor pressure, heat of vaporization, and the free volume, using the basic solubility equation and regarding vapor pressure as solubility in a medium in which the molecular forces are zero.

Answer: For a vacuum, $\delta_1 = 0$, and the actual volume occupied by gaseous molecules is, in general, small. Accordingly, ϕ_1 can be taken as 1. Thus, the change in enthalpy due to evaporation (i.e., solution in vacuum) is

$$V_2\phi_1^2(\delta_1 - \delta_2)^2 = V_2\delta_2^2 = \Delta E_2^V; \qquad \Delta E_2^V + RT = \Delta H_2^V$$

The liquid existing in equilibrium with the gas phase (the solution) is pure, and therefore the relative activity is equal to unity. Substituting these quantities into Eq. (4.10),

$$-R \ln p_2 c = \frac{\Delta H_2^V}{T}$$

where c is a conversion factor to permit the use of vapor pressure instead of the mole fraction x_2.

At constant pressure, for example, at the boiling points, ΔS_2^V is found to be essentially constant. That is to say, Trouton's rule can be derived from the solubility equation, if we assume $\Delta S_2^V = \Delta H_2^V/T \approx 21$ eu. Hence, $x_2 = p_2 c = 0.000028$.

At 25°C, 1 mol of gas occupies 24,400 cm^3 volume (free volume). If the free volume of one mole of liquid is taken as V_2^f, from Eq. (1.12),

$$\Delta S_2^V = R \ln \frac{V_2}{V_{2f}} = -R \ln x_2$$

Accordingly, we obtain $V_{2f} = 0.68$ cm^3. This value is in good agreement with known values of the free volume of liquids.

6.2 Calculate the change in the fugacity with temperature for the solvent in solution saturated with excess solid, under constant atmospheric pressure.

Answer: Let f_1 be the fugacity of the solvent in the solution, and f_2' and f_2 the fugacity of the solid and dissolved solute, respectively. As the fugacity of the solvent is a function of temperature and of composition,

$$d \ln f_1 = \frac{\partial \ln f_1}{\partial T} dT + \frac{\partial \ln f_1}{\partial x_1} dx_1$$

$$d \ln f_2 = \frac{\partial \ln f_2}{\partial T} dT + \frac{\partial \ln f_2}{\partial x_1} dx_1 = d \ln f_2' = \frac{\partial \ln f_2'}{\partial T} dT$$

Substituting for the term involving change of composition by means of the Gibbs-Duhem relation,

$$d \ln f_1 = \frac{\partial \ln f_1}{\partial T} dT + \frac{x_2}{x_1}\left(\frac{\partial \ln f_2}{\partial T} dT - \frac{\partial \ln f_2'}{\partial T} dT\right)$$

Making the further substitutions

$$\frac{\partial \ln f_1}{\partial T} \equiv \frac{H_1^* - \bar{H}_1}{RT^2}; \qquad \frac{\partial \ln f_2'}{\partial T} = \frac{H_2^* - H_2'}{RT^2}; \qquad \frac{\partial \ln f_2}{\partial T} = \frac{H_2^* - \bar{H}_2}{RT^2}$$

we obtain

$$\frac{\partial \ln f_1}{\partial T} \equiv \frac{H_1^* - \bar{H}_1}{RT^2} + \frac{x_2}{x_1}\frac{H_2' - \bar{H}_2}{RT^2}$$

where $H_1^* - \bar{H}_1$ is the molal (ideal) heat of vaporization of solvent, and $\bar{H}_2 - H_2'$ is the heat of solution of 1 mol of solute.

PROBLEMS

6.1 In solutions consisting of two components of equal molal volumes, the relation between the activity of a component and the concentration is given by

$$\ln a_2 = \ln x_2 + \frac{x_1^2 B}{RT}$$

Using this equation, derive an equation which describes the change in activity with concentration, and calculate the value of $(\partial \ln a_2/\partial \ln x_2)_{p,T}$ at $x_2 = 0, 0.2, 0.5$, and 1.0, when $B = 2RT$.

6.2 Examine the thermodynamic properties of an ideal solution, athermal solution, dilute solution, regular solution, associated solution, solvated solution, and polymer solution, in the table below, where component 1 is the solvent and component 2 is the solute. What conclusions can be drawn?

Type of solution	$\bar{H}_i - H_i^\circ$	$\bar{S}_i - S_i^\circ$	*Activity*
Ideal	0	$-R \ln x_1$	$a_1 = x_1$
Athermal	0	$> -R \ln x_1$	$a_1 < x_1$
Dilute	0	$-R \ln x_1$	$a_1 = x_1$
Dilute	+ or (~ -)	$-R \ln x_2$	$a_2 = kx_2$
Regular	+	$-R \ln x_1$	$a_1 > x_1$
Associated	+	$> -R \ln x_1$	$a_1 > x_1$
Solvated	-	$< -R \ln x_1$	$a_1 < x_1$
Polymer	≈ 0	$> -R \ln x_1$	$a_1 \approx x_1$
Polymer	+	$> -R \ln x_2$	$a_2 > x_2$

chapter seven

AQUEOUS SOLUTIONS

7.1 ORDER OUT OF CHAOS

In regular solutions liquid-liquid mixtures

$$\ln a_2 = \ln x_2 + \frac{V_2 \phi_1^2 B'}{RT} \qquad (4.10)$$

where x_2 is the solubility and $V_2\phi_1^2 B'$ is the enthalpy of solution, where $B' = (\delta_1 - \delta_2)^2$ may be determined experimentally. The relative activity a_2 is close to unity in a solution of a sparingly soluble liquid in another liquid. Since the logarithm of the activity would then be close to zero, the logarithm of the solubility and the energy of solution must just balance each other, i.e., $\ln x_2 \approx -V_2\phi_1^2(\delta_1 - \delta_2)^2/RT$. Hence, the enthalpy must be a large positive quantity for small x_2.

However, although the solubility of hydrocarbons, higher alcohols, and amines, etc., is very low, the heat of solution is generally small or even negative at room temperature. This deviation from what is predicted above has been ascribed to the formation of an "iceberg" (i.e., more ordered) structure in the water surrounding the solute molecule [1].

Systems in which water is a continuous phase are extremely important, since almost all biological systems are water-continuous, and, indeed, water is generally the most economical solvent. In an aqueous environment, the nonpolar groups of proteins tend to adhere to one another, and this tendency has been referred to as hydrophobic bonding. This terminology is inadequate, as no specific intermolecular bonding exists [2]. It has been concluded that hydrophobic bonding is caused by an entropy effect, and the solubility is decreased by the iceberg formation [3-6].

A thermodynamic analysis by the author shows, however, that

1. The small negative enthalpy of solution results from a large positive enthalpy of mixing and a large negative enthalpy of iceberg formation.
2. The large negative standard entropy of solution also results from iceberg formation.
3. The enthalpy decrease due to iceberg formation is largely cancelled by the accompanying entropy decrease (but the enthalpy is always the larger).
4. Iceberg formation is a phenomenon similar to phase change, which is neither an enthalpy nor entropy effect.
5. Iceberg formation increases the solubility [7].

7.2 ENTHALPY AND ENTROPY OF SOLUTION IN REGULAR SOLUTIONS

In order to understand the solution of hydrocarbons, nonelectrolytes, and so on, in water, it is useful to summarize the situation with respect to enthalpy and entropy change in regular solutions. The enthalpy and entropy change accompanying the solution of liquid component 2 (solute) in liquid component 1 (solvent) for regular solution is

$$\Delta S_2 = -R \ln x_2 \qquad \text{(pure liquid} \rightarrow \text{solute)} \qquad (7.1)$$

$$\Delta H_2 = V_2\phi_1{}^2B' \qquad \text{(pure liquid} \rightarrow \text{solute)} \qquad (7.2)$$

In the case where component 1 is sparingly soluble in component 2, the second component in the pure liquid state and in the solute state are equilibrium, so that the Gibbs free energy of component 2 does not change on solution, $\Delta G_2 = 0 = \Delta H_2 - T\Delta S_2$, and thus

$$V_2\phi_1^2B' = -RT \ln x_2 \tag{7.3}$$

If component 2, on the other hand, is a pure solid which does not form a solid solution with the solvent, the enthalpy and entropy of solution is given by the following:

$$\Delta H_2 = V_2\phi_1^2B' + \Delta H_2^F \qquad \text{(solid} \rightarrow \text{solute)} \tag{7.4}$$

$$\Delta S_2 = -R \ln x_2 + \Delta S_2^F \qquad \text{(solid} \rightarrow \text{solute)} \tag{7.5}$$

where ΔH_2^F and ΔS_2^F are the enthalpy and entropy of fusion of the solid at the temperature of solution. The solid melts at this temperature and mixes with the solvent in the process of solution. The additional term due to the heat of fusion is incidental to the solution of a solid in a liquid, as discussed in Chap. 6. Since the solid and solute states of component 2 are in equilibrium, we again have

$$\Delta G_2 = 0 = \Delta H_2 - T\Delta S_2 \tag{7.6}$$

Combining Eqs. (7.4), (7.5), and (7.6), we obtain

$$0 = \Delta H_2^F - T\Delta S_2^F + RT \ln x_2 + V_2\phi_1^2B' \tag{7.7}$$

If we represent the vapor pressure of the solid by p_2^s and that of the supercooled liquid as p_2^ℓ, the Gibbs free energy change of fusion is expressed as

$$\Delta G_2^F = \Delta H_2^F - T\Delta S_2^F = RT \ln\left(\frac{p_2^\ell}{p_2^s}\right) \tag{7.8}$$

Substituting Eq. (7.8) into (7.7),

$$RT \ln\left(\frac{p_2^s}{p_2^\ell}\right) = RT \ln a_2 = RT \ln x_2 + V_2\phi_1^2B' \tag{7.9}$$

The enthalpy of solution and the temperature dependence of solubility are related by

$$-R\left(\frac{\partial \ln a_2}{\partial \ln x_2}\right)_T\left(\frac{\partial \ln x_2}{\partial(1/T)}\right)_{sat} = \Delta\bar{H}_2 = T\Delta\bar{S}_2 \tag{7.10}$$

where $(\partial \ln a_2/\partial \ln x_2)_T \approx 1$ in dilute solutions.

It is clear from Eqs. (7.4) and (7.10) that the temperature dependence of the logarithm of the solubility is proportional to the enthalpy of solution, and increases discontinuously by the amount ΔH_2^F below the melting point of the solute, as shown in Fig. 7.1. However, the enthalpy and entropy of fusion cancel each other, so that solubility of solid and liquid are equal at the melting point.

In Fig. 7.1 we note that the solubility of the solid solute (S_2) becomes less than that of the supercooled liquid (dashed line) below the melting point of the solute. The ratio of the solubility of solid to that of the supercooled liquid is equal to the relative activity. As we shall see, if solvent molecules form an ordered structure around the solute, a phenomenon related to solidification and an inverse process to the solution of solid, the solubility of liquid solute will be greater than that in ordinary solution as shown by the broken curve in Fig. 7.1, and the heat of solution will decrease. These points are discussed in greater detail below.

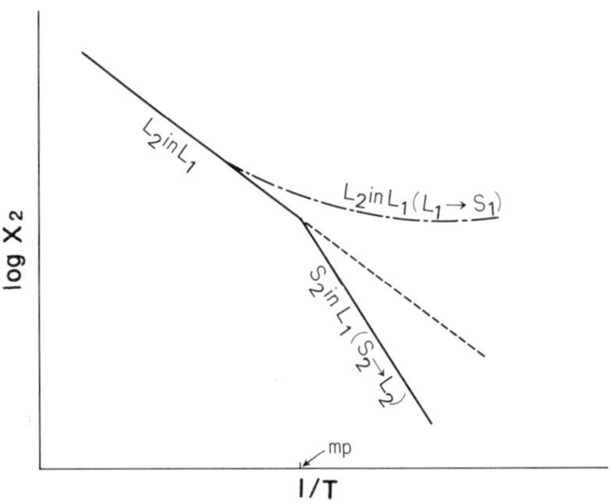

Fig. 7.1 Behavior of solid solubility above and below melting point.

7.3 ENTHALPY AND ENTROPY OF SOLUTION OF NONPOLAR SOLUTES IN WATER

In the process of solution of nonpolar solutes in water, iceberg formation and destruction of dipole-dipole interactions including hydrogen-bond deformation occur in the region around the nonpolar molecule. In dilute solutions, the amount of water which forms icebergs and in which hydrogen-bond bending takes place increases in proportion to the number of moles of solute, but does not increase with the amount of added solvent. Thus, any change in solute or solvent affects the partial molal quantity of solute, but has little effect on the solvent in dilute solution (see Sec. 1.6). This follows from the general partial molal equation for a binary system,

$$n_1 \, d\bar{Y}_1 + n_2 \, d\bar{Y}_2 = 0$$

where $\bar{Y}$ is any extensive property of the solution.

According to this equation, $d\bar{Y}_1$ is negligibly small in dilute solution, because n_2/n_1 is very small (about 10^{-3} in the system under consideration). If n moles of water form icebergs per mole of solute, the enthalpy and entropy of solution of nonpolar solutes in water is expressed as follows:

$$\Delta\bar{H}_2 = \Delta\bar{H}_2 \text{ (dipole)} + n \, \Delta\bar{H}_i \quad \text{(nonpolar liquid} \rightarrow \text{solute in } H_2O) \tag{7.11}$$

$$\Delta\bar{S}_2 = -R \ln x_2 + \Delta\bar{S}_2 \text{ (dipole)} + n \, \Delta\bar{S}_i \quad \text{(nonpolar liquid} \rightarrow \text{solute in } H_2O) \tag{7.12}$$

where $-n \, \Delta\bar{H}_i$ $(-n \, \Delta\bar{S}_i)$ is the partial molal enthalpy (entropy) of component 2 owing to iceberg formation of n moles of surrounding water, and $\Delta\bar{H}_2$ (dipole) [$\Delta\bar{S}_2$ (dipole)] is the partial molal enthalpy (entropy) change resulting from the bending or breaking of the hydrogen bond.

7.4 ENTHALPY OF MIXING OF HYDROCARBONS IN WATER

The surface tension γ of liquids is a direct measure of intermolecular forces. For saturated hydrocarbons, whose intermolecular forces are purely dispersion forces, γ_2 = 20-26 dyn/cm. For water, γ_1 = 72.8 dyn/cm, which is the sum of the contributions of dispersion forces and hydrogen-bonding interactions. The interfacial tension between water and saturated hydrocarbons, γ_{12}, is about 50-52 dyn/cm at 20°C. Introducing these values into Fowkes' equation, which relates the interfacial tension to the surface tensions of respective components,

$$\gamma_{12} = \gamma_1 + \gamma_2 - 2(\gamma_1{}^d \gamma_2)^{1/2} \tag{7.13}$$

one can evaluate the contribution of dispersion forces to the surface tension of water, which is $\gamma_1{}^d = 21.8 \pm 0.7$ [8]. This value is close to that of saturated hydrocarbons, i.e., the dispersion energy per unit volume of water is close to that of hydrocarbon. Hence, the enthalpy of mixing of hydrocarbons with water is mostly due to changes in the hydrogen-bonding interaction in the water surrounding the solute.

In the dissolution of hydrocarbons in water, the destruction of hydrogen-bonding interactions, and the formation of "icebergs" in the water surrounding the solute molecules may be the dominating processes. The enthalpy and entropy of solution of sparingly soluble hydrocarbons in water is expressed as follows, instead of by Eqs. (7.11) and (7.12),

$$\Delta\bar{H}_2 = \Delta\bar{H}_2(\text{H-bonds}) + n\ \Delta\bar{H}_i \tag{7.14}$$

$$\Delta\bar{S}_2 = -R \ln x_2 + \Delta\bar{S}_2(\text{H-bonds}) + n\ \Delta\bar{S}_i \tag{7.15}$$

where $\Delta\bar{H}_2$(H-bonds) [$\Delta\bar{S}_2$(H-bonds)] is the partial molal enthalpy (entropy) of the solute owing to the decrease in hydrogen-bonding interactions, and $-R \ln x_2$ is the entropy of dilution.

The extent of iceberg formation of the solvent (as shown by the value of n) will diminish with rise of temperature, and the enthalpy of solution, $-R\ \partial \ln x_2/\partial(1/T)$, should approach a more constant value. The enthalpy of solution at the temperature at which the iceberg formation disappears (approximately 160°C) may be evaluated as follows:

$$-R\,\frac{\partial \ln x_2}{\partial(1/T)} = \Delta\bar{H}_2 = \Delta\bar{H}_2(\text{H-bonds}) \qquad \text{at } 160°\text{C} \tag{7.16}$$

$$\frac{\Delta\bar{H}_2}{T} = \Delta\bar{S}_2 = -R \ln x_2 + \Delta\bar{S}_2(\text{H-bonds}) \qquad \text{at } 160°\text{C} \tag{7.17}$$

Once $\Delta\bar{H}_2$(H-bonds) and $\Delta\bar{S}_2$(H-bonds) are determined, $n\ \Delta H_i$ or $n\ \Delta S_i$ may be evaluated at any temperature from the solubility and the temperature dependence of the solubility by the aid of Eqs. (7.14) and (7.15).

7.5 ANALYSIS OF THE TEMPERATURE DEPENDENCE OF THE SOLUBILITY OF HYDROCARBONS IN WATER

The solubilities of liquid aromatic hydrocarbons are known up to higher temperatures [9,10]. The logarithms of solubilities versus the reciprocal of temperature are plotted in Fig. 7.2.

As the iceberg formation decreases with temperature rise, the slope, $-\partial \ln x_2/\partial(1/T)$, increases and becomes almost constant at approximately 160°C, at which temperatures iceberg formation seems mostly to disappear. The straight lines in Fig. 7.2 express the hypothetical solubility of the respective hydrocarbons in water assuming no iceberg formation. The solubility is much larger at lower temperature, as compared with the hypothetical solubility assuming no iceberg formation.

The solubilities of alkanes (C_1 to C_4) in water at one atmosphere as a function of temperature is known also [11-13]. These values were recalculated to the solubility of hypothetical liquid alkanes by the aid of an empirical curve, i.e., the relation of the fugacity of liquid divided by critical pressure to the reduced temperature, obtained by Prausnitz on the basis of the principle of corresponding states [14]. The extrapolated solubilities of liquid alkanes at higher temperatures show similar behavior to those of aromatic hydrocarbons.

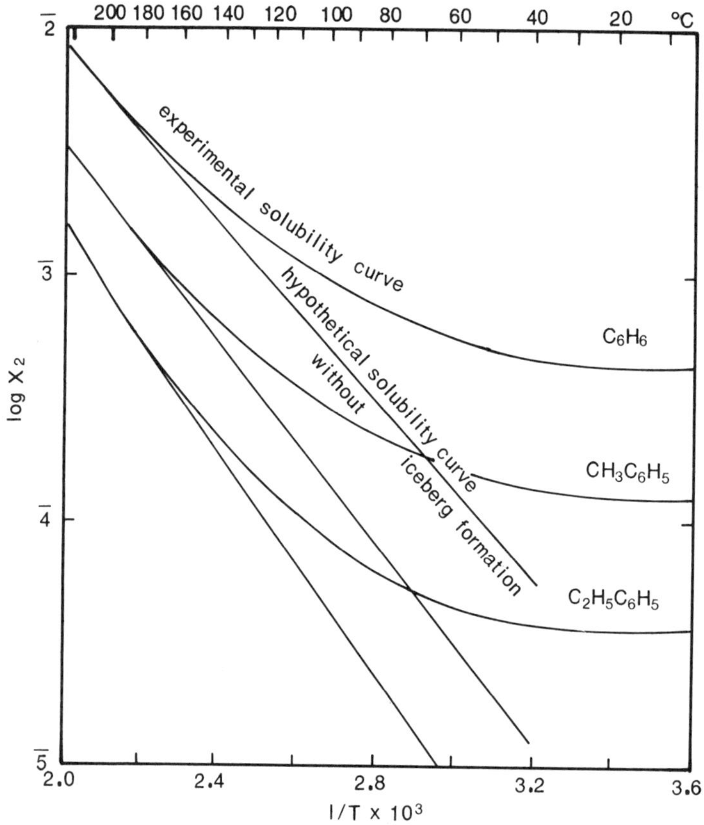

Fig. 7.2 Solubility of liquid hydrocarbons in water as a function of temperature.

The enthalpy and entropy of solution of liquid hydrocarbons in water at 25°C are calculated from the temperature dependence of the solubility and shown in Table 7.1.

The enthalpy of solution $\Delta\bar{H}_2$(H-bonds) is obtained from the slope of the straight lines in Fig. 7.2, then the enthalpy change due to the iceberg formation $n\ \Delta\bar{H}_i$ is readily obtained by the aid of Eq. (7.14). The entropy change due to the hydrogen-bonding interactions $\Delta\bar{S}_2$(H-bonds) is evaluated by the aid of Eq. (7.17), and $n\ \Delta\bar{S}_i$ is then obtained from Eq. (7.15). All these values are summarized in Table 7.1. As the solubilities of water in these

Table 7.1 The Crude Analysis of the Enthalpy and Entropy of Solution of Liquid Hydrocarbons in Water at 25°C (First Approximation)[a]

Solute	$\Delta\bar{H}_2 = \Delta\bar{H}_2(\text{H-bonds}) + n\,\Delta\bar{H}_i$ (cal/mol)			$\Delta\bar{S}_2 = -R \ln x_2 + \Delta\bar{S}_2(\text{H-bonds}) + n\,\Delta\bar{S}_i$ (cal/deg/mol)				$\bar{T}_i = \Delta\bar{H}_i/\Delta\bar{S}_i$ (K)	V_2 (cm^3)
CH_4	-1900 =	6700	- 8600	-6.4 =	10.8	+ 5.6	- 22.8	379	--
C_2H_6	-2000 =	7900	- 9900	-6.7 =	13.7	+ 6.5	- 26.9	368	--
C_3H_8	-1900 =	9500	- 11400	-6.3 =	16.5	+ 7.9	- 30.7	372	89
C_4H_{10}	-800 =	11200	- 12000	-2.7 =	19.6	+ 9.0	- 31.3	383	101
C_6H_6	+480 =	8100	- 7520	+1.9 =	15.4	+ 7.2	- 20.7	363	89
$CH_3C_6H_5$	+640 =	9260	- 8620	+2.2 =	17.9	+ 7.9	- 23.6	365	107
$C_2H_5C_6H_5$	+400 =	10500	- 10100	+1.4 =	20.4	+ 8.4	- 27.4	370	123

[a]Enthalpy of solution = $\Delta\bar{H}_2 = -R(\partial \ln x_2/\partial(1/T))_{sat}$; entropy of solution = $\Delta\bar{H}_2/T$; enthalpy [entropy] of mixing (H-bonds interaction) = $\Delta\bar{H}_2(\text{H-bonds})[\Delta\bar{S}_2(\text{H-bonds})]$; entropy of dilution = $-R \ln x_2$; enthalpy of iceberg formation = $n\,\Delta\bar{H}_i$; entropy of iceberg formation = $n\,\Delta\bar{S}_i$; mean temperature of iceberg formation = $\bar{T}_i = \Delta\bar{H}_i/\Delta\bar{S}_i$; molar volume at 25°C = V_2.

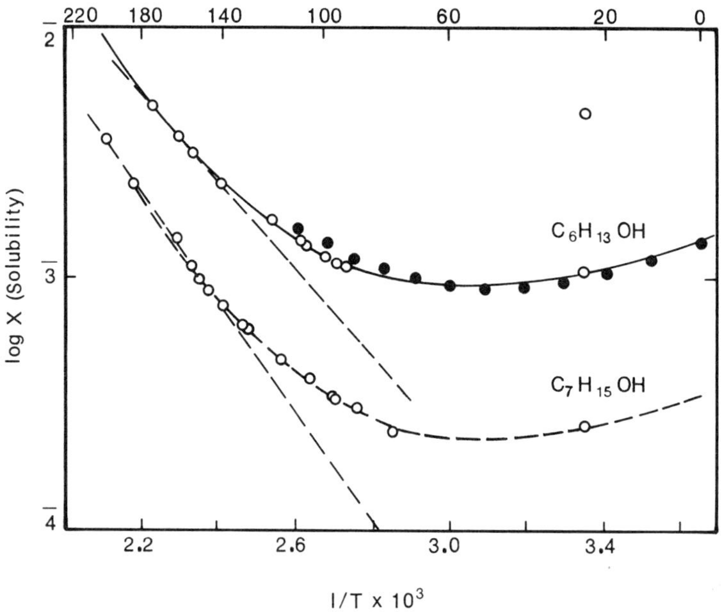

Fig. 7.3 Solubility of alkanols in water as a function of temperature.

hydrocarbons are so small, $x \approx 10^{-3}$, the separated hydrocarbon phase can be assumed to be a pure phase [15].

The enthalpy of mixing is approximately proportional to the molal volumes of respective alkanes, and the ratio $\Delta\bar{H}_i/\Delta\bar{S}_i$ = 360 to 380 K, which is the mean temperature of iceberg formation analyzed at 25°C, is nearly constant for all hydrocarbons examined. As iceberg formation occurs above room temperature, this is a reasonable first approximation. The amount of the enthalpy and entropy loss due to the iceberg formation decreases with temperature rise and the ratio of these values is enhanced, as would be expected. It is evident from Fig. 7.2 and Table 7.1 that the enthalpy of mixing has a large positive value, as would be expected on the basis of the intermolecular forces. Small negative enthalpy of solution $\Delta\bar{H}_2$ results from a large positive enthalpy of mixing and a large negative enthalpy of iceberg formation. Table 7.1 and Fig. 7.2

enable us to understand that the small solubility of hydrocarbons in water is mainly the result of the large positive enthalpy of mixing, and that the enthalpy decrease of the iceberg formation is largely cancelled [16,17] by the accompanying entropy decrease at 25°C. The solubility increase from the solid line to the experimental curve in Fig. 7.2 is caused by iceberg formation.

Similar behavior is observed in the solution of rare gases, oxygen, nitrogen, as well as other organic substances, such as alcohols, as shown in Fig. 7.3.

REFERENCES

1. H. S. Frank and M. W. Evans, J. Chem. Phys., *13*, 507 (1945).
2. J. H. Hildebrand, J. Phys. Chem., *72*, 1841 (1968); K. Shinoda, Chem. Chem. Ind. (Japan), *21*, 1401 (1968) (in Japanese).
3. I. M. Klotz, Science, *128*, 815 (1958).
4. W. Kauzmann, Adv. Protein Chem., *14*, 37 (1959).
5. G. Nemethy and H. A. Scheraga, J. Chem. Phys., *36*, 3382 (1962); *36*, 340 (1962).
6. G. Nemethy and H. A. Scheraga, J. Phys. Chem., *66*, 1173 (1962).
7. K. Shinoda, J. Phys. Chem., *81*, 1300 (1977).
8. F. M. Fowkes, J. Phys. Chem., *67*, 2538 (1963).
9. *API Technical Data Book--Petroleum Refining*, 2d ed., American Petroleum Institute, 1970.
10. A. N. Guseva and E. I. Parnov, Vestn. Mosk. Univ. Ser II, Khim., *18*, 76 (1963); C.A. *58*, 9673f; D. M. Alexander, J. Phys. Chem., *63*, 1021 (1959).
11. W. F. Claussen and M. F. Polglase, J. Am. Chem. Soc., *74*, 4817 (1952).
12. T. J. Morrison and F. Billet, J. Chem. Soc., *1952*, 3819.
13. L. W. Winkler, Ber., *36*, 1408 (1901).
14. J. M. Prausnitz, J. Phys. Chem., *66*, 640 (1962).
15. C. Black, G. G. Joris, and H. S. Taylor, J. Chem. Phys., *16*, 537 (1948).
16. H. S. Frank, Fed. Proc. Suppl., *24*, 1 (1965).
17. D. Patterson and M. Barbe, J. Phys. Chem., *80*, 2435 (1976).

chapter eight

POLYMER SOLUTIONS

8.1 INTRODUCTION

The well-known book of Flory [1] and others may be consulted for discussions of the general physical-chemical properties of polymer solutions. However, the discussion of the solubility of polymers is limited. In this chapter, the solubility and the dissolution state of polymers including water-soluble polymers are described. There are at least two different types of polymer solutions. One is a regular polymer solution, such as rubber in hydrocarbon, in which solute and solvent molecules randomly mix by thermal motion. The other is a nonregular polymer solution, such as biocolloids in water, in which the orientation, arrangement, conformation, and structure formation due to polymer molecules are important. The two types of polymer solutions are quite different. Nevertheless, these solutions exhibit common characteristic properties of polymer solution, as follows:

1. Because of the large intermolecular forces between polymer molecules they exist as solids, liquids, or solutions, but not gases because their equilibrium vapor pressure is zero. That is, solubility in a medium in which the molecular forces are very different (including vacuum) is practically zero.

2. When the polymer molecules are oriented in their long dimension, so that the molecular forces are cumulative, the mechanical

strength is increased to an extent which makes possible the forming of the polymer into filaments and films. In this situation, the solution viscosity becomes very high.

3. The polymers may show rubberlike elasticity over a range of temperatures because the long-chain molecules may arrange in a large number of different configurations. The elasticity is thus caused by an entropy effect, absent in low-molecular-weight substances.

4. Polymers are usually very slowly crystallized. When polymer in the liquid state is cooled it solidifies in a glassy state. Thus, plastics are usually in a glassy state and show plasticity.

5. Polymers with a three-dimensional network structure do not dissolve; however, solvent penetrates the polymer, causing swelling.

6. If the polymer concentration is greater than about 5% it may be assumed that the solvent is dissolved in the polymer as the solute, even for linear polymers. This is demonstrated by gel formation in 1-3% solutions of agar or starch.

If the polymer and solvent are mutually soluble, the heat of solution per unit volume is small. A feature of polymer solution is that the entropy of mixing is large compared to that for ordinary solutions. Mole fraction is not a satisfactory measure of composition for high polymer solutions, since the molecular volume of the solute is so greatly different from that of the solvent. Namely, the molecular weight of the polymer is not accurately known on the one hand, and the mole fraction of solvent is nearly unity in almost all cases on the other hand. Weight fraction is a convenient measure, since it is not dependent on temperature or pressure. As a practical matter, however, the volume fraction of the respective components before mixing (at the same temperature) is customarily used.

The activity of benzene in rubber-benzene solutions has been obtained from the osmotic pressure of the solution and the partial pressure of benzene. The activity of benzene as a function of the composition is shown in Fig. 8.1 [2].

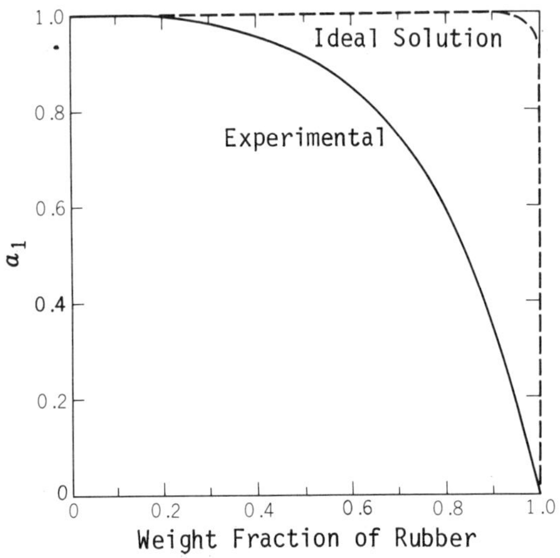

Fig. 8.1 Activity of benzene in rubber-benzene solutions compared with theoretical prediction for an ideal solution of a solute with molecular weight 280,000 [2].

Since the molecular weight of rubber is so much larger than that of benzene, the slope of the activity or entropy versus the weight fraction of rubber is almost zero in the region of low concentration, as can be seen in Fig. 8.1. This observation coincides well with the law of dilute solutions. However, in concentrated solutions, a large discrepancy between the observed and ideal curves is found.

In Chap. 6 we saw that the entropy of mixing is almost equal to that of an ideal solution, when the ratio of the molal volumes is about 1:8 in regular solutions. However, since the polymer molecule is much larger than that of the solvent, and since it is a long chain, the free volume of the molecule becomes significantly larger, and thus there results the large discrepancy between the observed and ideal cases. This phenomenon can be explained by consideration of the large number of configurations available to the molecule, and to its freedom of movement.

8.2 ENTROPY OF MIXING ACCORDING TO THE LATTICE MODEL

Flory calculated the number of configurations available by arranging N_1 solvent molecules and N_2 chainlike molecules containing x segments on $N = N_1 + xN_2$ lattice points (Fig. 8.2). The quantity x is the ratio of the molal volumes of the solute and solvent. A segment and a solvent molecule may thus replace one another in the lattice. First, considering the lattice points to be vacant, we place the x segments, belonging to the first chain molecule, on the lattice points. This is repeated, one by one, for all the N_2 molecules, and then the N_1 solvent molecules are placed on the remaining lattice sites. The configurations corresponding to the placement are calculated by noting that the number of vacant lattice sites left after N_2' chain molecules have been placed is equal to $N - xN_2'$. Hence, the number of ways of placing the first segment of the $(N_2' + 1)$st molecule equals $N - xN_2'$. Taking z as the coordination number (number of neighbors) of the lattice, the number of ways of arranging the second segment of the $(N_2' + 1)$st molecule may be taken as $z(N - xN_2')/N$ (the assumption being that the local density of vacant sites is equal to the average density of the

Fig. 8.2 Lattice model for polymer solution.

whole lattice). The expected number of lattice sites available to the third segment will be (z - 1), since one of the sites adjacent to the second segment is occupied by the first. Therefore, the number of sites available to the third segment will be $(z - 1)(N - xN_2')/N$. It is further assumed that the number of configurations is the same for each succeeding segment, although it ought to be less than for the third.

Accordingly, the number of configurations for the $(N_2' + 1)$st chain molecule is

$$W_{N_2'+1} = \frac{1}{2}(N - xN_2')\frac{z(N - xN_2')}{N}\left[(z - 1)\frac{(N - xN_2')}{N}\right]^{x-2}$$

$$\doteqdot \frac{1}{2}\frac{(z - 1)^{x-1}}{N^{x-1}}(N - xN_2')^x \tag{8.1}$$

The factor of 1/2 in the above equation arises from the symmetrical nature of the chain molecule. However, the factor 1/2 is eliminated in the calculation of entropy of mixing, so that one obtains the same result for unsymmetrical molecules. The number of arrangements for all the chainlike polymer is

$$W = \frac{1}{N_2'}\prod_{N_2'=0}^{N_2-1} W_{N_2'} + 1 \tag{8.2}$$

from which

$$W = \left(\frac{z - 1}{N}\right)^{(x-1)N_2} \frac{x^{xN_2}}{2^{N_2}N_2!}\left[\frac{(N/x)!}{(N/x - N_2)!}\right]^x \tag{8.3}$$

The quantity $N_2!$ is introduced in Eq. (8.3) to take into account the interchangeability of the N_2 polymer molecules, that is, the N_2 may be permuted without changing the state of the solution.

Using Stirling's approximation

$$W = \left(\frac{z - 1}{e}\right)^{(x-1)N_2} \frac{1}{2^{N_2}} \frac{(N_1 + N_2)^{N_1+N_2}}{N_1^{N_1} N_2^{N_2}} \tag{8.4}$$

Since the number of configurations obtained by placing the N_1 solvent molecules into the remaining lattice sites is $N_1!/N_1! = 1$, the value of W is not affected.

Thus the entropy may be calculated by applying Boltzmann's equation, $S = k \ln W$,

$$\Delta S_{M'} = -k\, N_1\left[\ln \frac{N_1}{N_1 + xN_2} + N_2 \ln \frac{N_2}{N_1 + xN_2}\right] + kN_2(x - 1)[\ln(z - 1) - 1] - kN_2 \ln 2 \tag{8.5}$$

where $\Delta S_{M'}$ is the configurational entropy of the solution. The entropy of solution may be calculated by deducing the configurational entropy of the chainlike polymer in the absence of solvent. Although the assumption is far less adequate than those used to obtain Eq. (8.5), we can obtain the entropy of configuration of the polymer per se by substituting $N_1 = 0$ in Eq. (8.5),

$$\Delta S_p = kN_2\{\ln x/2 + (x - 1)[\ln(z - 1) - 1]\} \tag{8.6}$$

Taking the difference between Eqs. (8.5) and (8.6), we obtain

$$\Delta S_M = k\left[N_1 \ln \frac{N_1}{N_1 + xN_2} + N_2 \ln \frac{xN_2}{N_1 + xN_2}\right] \tag{8.7}$$

It is interesting to note that the mole fractions which occur in the expression for ideal entropy of mixing are replaced by volume fraction in the entropy of mixing of molecules of different sizes, i.e.,

$$\Delta S_M = -k(N_1 \ln \phi_1 + N_2 \ln \phi_2) \tag{8.8}$$

Replacing N_1 and N_2 by n_1 and n_2 (moles), and k by R, and then differentiating with respect to n_2,

$$\Delta \bar{S}_2 = \left(\frac{\partial \Delta S_M}{\partial n_2}\right)_{n_1} = -R[\ln \phi_2 + \phi_1(1 - x)] \tag{8.9}$$

and since $x = V_2/V_1$

$$\Delta\bar{S}_2 = -R\left[\ln \phi_2 + \phi_1\left(1 - \frac{V_2}{V_1}\right)\right] \tag{8.10}$$

and

$$\Delta\bar{S}_1 = -R\left[\ln \phi_1 + \phi_2\left(1 - \frac{V_1}{V_2}\right)\right] \tag{8.11}$$

This is the result obtained by Flory [2]. Huggins [4], Miller [5] and Guggenheim [6] have obtained similar relations. Equations (8.10) and (8.11) comprise the so-called Flory-Huggins entropy.

In multicomponent systems consisting of polymers of different molecular weights, the following equation applies:

$$\Delta S_M = -k \sum N_i \ln \phi_i \tag{8.12}$$

Before we proceed with the application of the Flory-Huggins entropy, it is desirable to examine the integrated approximations involved in the derivation. It will be observed that the term containing the number of neighboring sites z is eliminated in Eqs. (8.8) to (8.12), which suggest that these equations might be derived on the basis of models other than the lattice. As a matter of fact, Hildebrand [7] has shown that the Flory-Huggins entropy of mixing may be derived assuming that the ratio of the free volumes of solvent to polymer is equal to the ratio of their molal volumes. This assumption is equivalent to picking x unconnected lattice points and regarding them as being one polymer molecule, or to assuming complete flexibility, expandability, and contractability of the polymer molecule, i.e., to assume $z = \infty$, or, in other words, that the free volume is proportional to the molal volume. However, such an assumption is surely an overestimate of the flexibility on the one hand, and unrealistic, since bond length scarcely changes, on the other hand. In addition, as stated in Chap. 1, the free volume of a liquid is not proportional to the molal volume. The entropy of configuration calculated from the Flory-Huggins equation must be an overestimate (see Chap. 6). However, the excess entropy of configuration, or the difference in the free volumes of the components, will give a reasonable explanation of the behavior of polymer solutions. If we could rigorously estimate the correct configurational entropy, or could estimate the respective free volumes, or the ratio of the free volumes of the components, and replace volume fraction with free volume fraction in the entropy term, the revised Eqs. (8.8) to (8.12) would agree much better with experiment.

The partial molal entropies for the systems benzene-rubber, poly(dimethyl siloxane)-benzene, polystyrene-methyl ethyl ketone (•), and polystyrene-toluene (○) are shown in Fig. 8.3 [3].

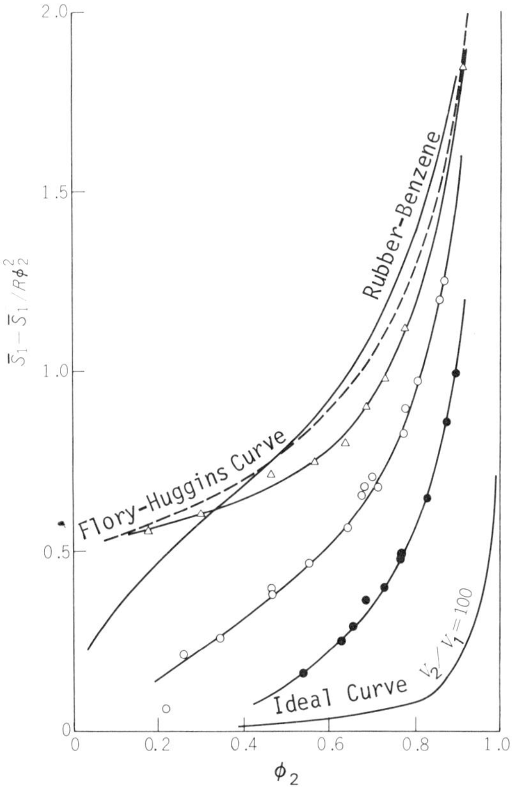

Fig. 8.3 Comparison of observed values of partial molal entropy of the solvent with the predictions of the Flory-Huggins and ideal entropy curves [3].

The values calculated from the Flory-Huggins equation (broken line), and the theoretical value for ideal mixing for a solution of solvent and polymer in the volume ratio 1:100 are also plotted for comparison.

It is seen readily that the calculated curves do not agree well with the observed values, the deviation being in the direction of ideal mixing. In the more detailed treatment the correction for the entropy is included as the entropy part of the interaction parameter [8].

8.3 SEPARATION INTO TWO PHASES AND MUTUAL SOLUBILITY

Just as for regular solutions, the heat of mixing of polymer solutions is given by

$$\Delta\bar{H}_1 = B'V_1\phi_2^{\,2}; \qquad \Delta\bar{H}_2 = B'V_2\phi_1^{\,2} \tag{8.13}$$

Combining Eqs. (8.11) and (8.13), the Gibbs energy is

$$\Delta\bar{G}_1 = RT\left[\ln \phi_1 + \phi_2\left(1 - \frac{1}{x}\right) + \frac{B'V_1}{RT}\,\phi_2^{\,2}\right] \tag{8.14}$$

Thus, the vapor pressure becomes

$$\ln \frac{p_1}{p_1^\circ} = \ln \phi_1 + \phi_2\left(1 - \frac{1}{x}\right) + \frac{B'V_1}{RT}\,\phi_2^{\,2} \tag{8.15}$$

If $\partial\Delta\bar{G}_1/\partial x_1 > 0$ at any concentration the polymer may dissolve completely in the solvent. As the heat of solution $B'V_1$ increases, the relative activity increases, and, at a given value of the solution, separation into two phases occurs.

The relation between the partial molal Gibbs free energy of solution and the heat of solution, $\mu = B'V_1/RT$, in polymer solution is plotted in Fig. 8.4 in the case when $V_2/V_1 = 100$. If $B'V_1 > 0.605$ RT, the solution splits into two phases.

Applying the critical conditions, the first and second derivatives of $\Delta\bar{G}_1$ with respect to composition are set equal to zero

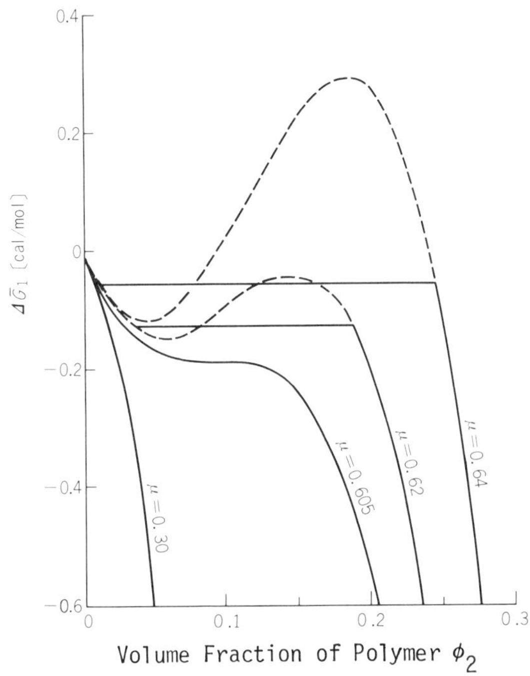

Fig. 8.4 Relation between the partial molal Gibbs free energy of solution of the solvent and the heat of solution ($\mu = B'V_1/RT$) for polymer solution. $V_2/V_1 = 100$.

$$\frac{\partial \Delta \bar{G}_1}{\partial \phi_1} = -RT_c\left[\frac{1}{\phi_{1c}} - \left(1 - \frac{1}{x}\right) - \frac{2B'V_1}{RT_c}\phi_{2c}\right] = 0 \tag{8.16}$$

$$\frac{\partial^2 \Delta \bar{G}_1}{\partial \phi_1{}^2} = -RT_c\left[-\frac{1}{\phi_{1c}{}^2} + \frac{2B'V_1}{RT_c}\right] = 0 \tag{8.17}$$

Eliminating ϕ_{1c} and ϕ_{2c}

$$\frac{2B'V_1}{RT_c} = \left(1 \pm \frac{1}{x^{1/2}}\right)^2$$

Since $\phi_{1c} < 1$ in Eq. (8.17), $2B'V_1/RT_c$ must be larger than one

$$\frac{2B'V_1}{RT_c} = \left(1 + \frac{1}{x^{1/2}}\right)^2 \tag{8.18}$$

Substituting Eq. (8.18) into Eq. (8.17) to eliminate $B'V_1/RT$,

$$\frac{x_{1c}}{x_{2c}} = \left(\frac{V_2}{V_1}\right)^{3/2}; \qquad \frac{\phi_{1c}}{\phi_{2c}} = \left(\frac{V_2}{V_1}\right)^{1/2} \tag{8.19}$$

Thus, we obtain

$$\phi_{2c} = \frac{1}{1 + x^{1/2}}; \qquad x_{2c} = \frac{1}{(V_2/V_1)^{3/2} + 1} \tag{8.20}$$

$$\phi_{1c} = \frac{x^{1/2}}{1 + x^{1/2}}; \qquad x_{1c} = \frac{(V_2/V_1)^{3/2}}{(V_2/V_1)^{3/2} + 1} \tag{8.21}$$

where $x = V_2/V_1$.

As is apparent from Eq. (8.18), $B'V_1 = 2RT_c$ when the molal volumes of the solvent and solute are equal, which agrees with the case of regular solutions. $B'V_1 = 0.605RT_c$ when $x = 100$, and $B'V_1 = 0.5RT$ when $x = \infty$.

This implies that the heat of solution per solvent molecule must be less than one-fourth that of a regular solution in order to be completely mutually soluble. Equation (8.20) implies that the critical composition for the liquid-liquid mixture is $\phi_2 \approx 0$ and $\phi_1 \approx 1$, when the molecular weight of the polymer is very high. From Eq. (8.20) we can conclude that the solvent phase in equilibrium with the polymer phase contains a very small amount of polymer. The polymer phase, however, has a considerable amount of solvent dissolved in it.

The mutual solubility curves for poly(isobutylene) at three different degrees of polymerization in diisobutyl ketone are shown in Fig. 8.5 [9].

The observed values of critical composition for the liquid-liquid solubility curve are not as small as the calculated ones, but lie closer to the values for the solvent. The broken line in Fig. 8.5 shows the calculated values.

If a polymer in equilibrium with a solvent is in the crystalline form, it generally does not dissolve, because the enthalpy of

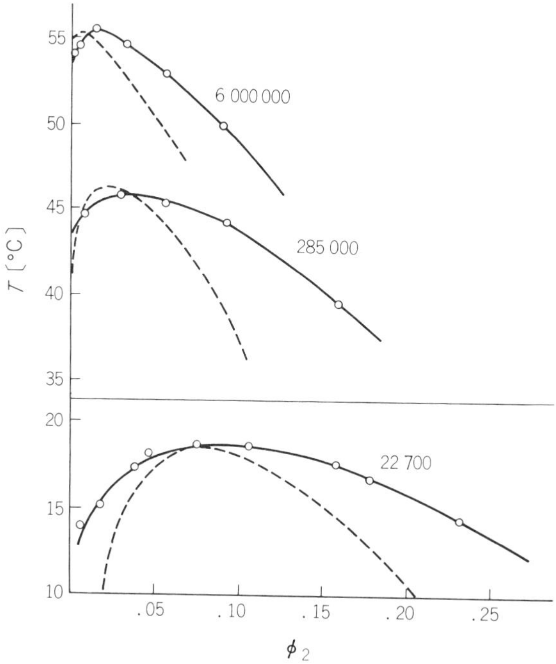

Fig. 8.5 Phase diagram for three poly(isobutylene) fractions in diisobutyl ketone. Solid lines are experimental data for the respective molecular weights; dashed lines are calculated from theory [9].

fusion of polymer is so large that the relative activity of solid in Eq. (4.10) decreases very rapidly with decreasing temperature. If the polymer is in the liquid state, the solvent molecules dissolve in the polymer. As the molal volume of the solvent is much smaller than that of the polymer, the heat of solution of the solvent is small, and it therefore dissolves readily. This explains the fact that when the two components are not completely soluble, the ratio of solvent to polymer even in polymer phase is so large that two phases form. In that situation, the heat of transfer of polymer from the polymer phase to the solvent phase becomes small. Namely, in the case of mutual solubility of solvent and polymer both ϕ_1 and ϕ_1' are close to one in the following equation:

$$\ln a_2 = \ln x_2 + \frac{B'V_2\phi_1{}^2}{RT} = \ln x_2' + \frac{B'V_2\phi_1'^2}{RT} \tag{4.10}$$

whereas in ordinary solutions $\phi_1 \approx 1$ and $\phi_1' \approx 0$. Here ϕ_1' is the volume fraction of the solvent in the solute (second component).

When the proper solvent is chosen, a large amount of solvent may dissolve in the polymer phase, resulting in a swelling of the polymer phase, and then $\phi_1 \approx 1.0\text{-}0.9$. Rearranging Eq. (4.10), we obtain

$$\ln x_2' = \ln x_2 + \frac{B'V_2[\phi_1{}^2 - (\phi_1')^2]}{RT} \tag{8.22}$$

Thus, it is apparent that the difference between the heats of solution in the two phases is much smaller. The solubility of the solvent (component one) becomes

$$\ln a_1 = \ln x_1 + \frac{B'V_1\phi_2{}^2}{RT} = \ln x_1' + \frac{B'V_1(\phi_2')^2}{RT} \tag{8.23}$$

where $V_1 << V_2$, and therefore the amount of polymer dissolved in the solvent phase is small, even though the solvent dissolves well in the polymer phase, so that $\phi_2 \approx 0$, $x_1 \approx 1$, and $\ln a_1 \approx 0$. Accordingly, the amount of solvent dissolved in the polymer phase is expressed as

$$\ln x_1' + \frac{B'V_1(\phi_2')^2}{RT} = 0 \tag{8.24}$$

This means that when the solvent dissolved infinitely in the polymer phase, we may obtain solutions of any composition whatsoever. It is apparent from Eq. (8.18) that the small heat of solution of solvent in polymer,

$$B'V_1 = \frac{1}{2}\left(1 + \frac{1}{x^{1/2}}\right)^2 RT_c \tag{8.18'}$$

is the necessary condition for complete dissolution. For this purpose, the selection of a solvent with a small heat of solution per unit volume and a small molal volume is necessary. On the contrary,

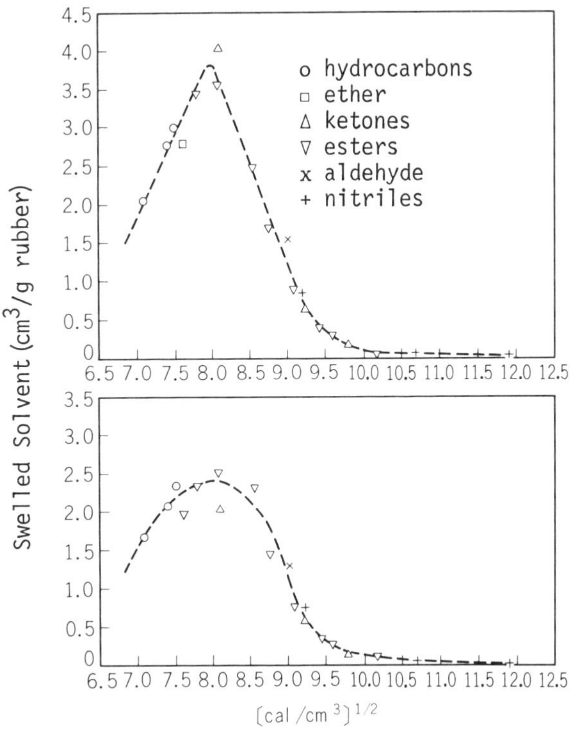

Fig. 8.6 Swelling of rubber in various solvents. Top, pure gum; bottom, tire tread.

the limiting allowable heat of solution is not much affected by the molal volume of the polymer. For example, when V_2/V_1 changes from 100 to infinity, $B'V_1$ changes only from $0.605RT_c$ to $0.5RT_c$. Hence, solution of polymers means essentially solution of solvent in the polymer. It is necessary that the polymer be in the liquid state, and that the solvent be readily soluble in the polymer. If the polymer has a three-dimensional network structure, the better is the solvent, the more swelling results, but the polymer does not dissolve into the solvent. In other words, the swelling ceases at a certain stage.

The solubility parameter is a useful parameter for the choice of a solvent readily soluble in the polymer. By the same token, when a good solvent is found, the required solubility parameter of the polymer may be determined. However, the accuracy is not high. Thus, the solubility parameter should be used only as a guide in the selection of a good solvent.

The relation between the swelling of natural rubber and the solubility parameters of various solvents is shown in Fig. 8.6. By this method, solubility parameters can be determined with a precision of ±0.4 units. Such parameters are tabulated in Table 8.1 [11-13].

Table 8.1 Solubility Parameters of Polymers

Polymers	δ	Reference
Polytetrafluoroethylene	6.2	a
Polydimethylsilicone	7.3	a
Polyethylene	7.87	b
Polyisobutylene	8.05	b
Polybutadiene	8.40-8.60	b
Polystyrene	9.1	b
Polyvinylchloride	9.5-9.7	b, a
Polyvinylacetate	9.4	a
Polymethylmethacrylate	9.5	a
Polyglycol terephthalate	10.7	a
Polymethacrylonitrile	10.7	a
Cellulose diacetate	10.9	a
Nylon 66	13.6	a
Polyacrylonitrile	15.4	a

[a] H. Burrell, Interchem. Rev., *14*(3), 31 (1955).

[b] H. Mark and A. V. Tobolsky, *Physical Chemistry of High Polymeric Systems*, Interscience, New York, 1950, p. 263.

8.4 SOLUBILITY OF POLYELECTROLYTES [14]

Polymers with dissociable groups are called polyelectrolytes. For example, such natural products are alginic acid, gum arabic, agar, gelatin (protein), nucleic acid, derivatives of natural products such as viscose and CMC (carboxymethylcellulose), and synthetic substances such as poly(acrylic acid) $\{-CH\ COOH-CH_2\}_x$, poly(methacrylic acid) $\{-C(CH_3)COOH-CH_2\}_x$, the sodium salt of poly(styrene) sulfonic acid $\{-CH_2CHC_6H_5SO_3Na\}_x$, poly(vinyl pyridine) $\{-CH_2CHC_5H_5N\}_x$. Ion-exchange resins are an example of a polyelectrolyte possessing a three-dimensional network structure.

A polyelectrolyte is a substance which can be described as a random-coil model of a long-chain polymer existing in solution much in the manner indicated in Fig. 8.7. Polyelectrolytes are readily dissolved in solvents in which regular polymers are difficult to dissolve. This is due to the fact that the dissociation energy in a solvent with a high dielectric constant, such as water, decreases inversely with the dielectric constant, while the large entropy increase due to dissociation does not depend on the dielectric constant, and thus promotes the dissociation. As the dissociated ions dissolve in the solvent, the remaining polyelectrolyte molecules begin to bear electric charges, resulting in electric repulsion within

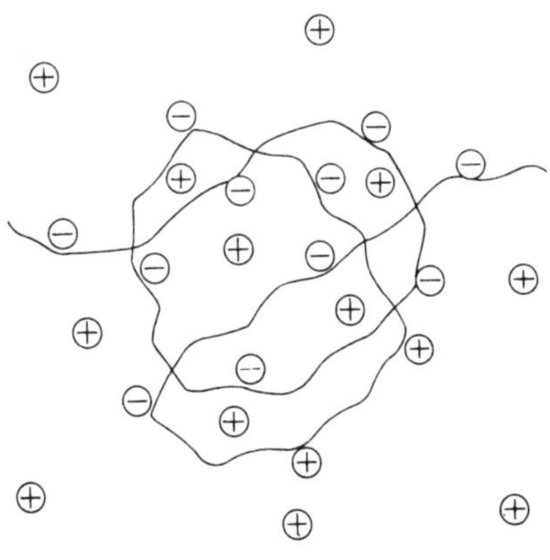

Fig. 8.7 Schematic representation of polyelectrolyte dissolved in water.

the molecules and between the molecules, while solvation of the dissociated ions further increases the affinity toward the solvent (i.e., decreasing heat of solution). Thus, the random coil of the polymer is thought to be much larger as a result of the dissociation, resulting in better solubility. In the case of low polyelectrolyte concentration and in the absence of salts, the counterions dissociate with a larger entropy increase, resulting in expansion of the molecule as the polyelectrolyte molecules elongate themselves. When the dissociated groups are densely concentrated along the chain, the polyelectrolyte molecule is elongated, because of repulsion between the groups.

On the other hand, the addition of multivalent counterions, even in small amount, weakens this tendency greatly, resulting in precipitation due to loss of solubility. This is generally due to the fact that as the electrical repulsion forces decrease, the dissociation decreases, and on the other hand, the melting point of the multivalent salt of the polyelectrolyte is enhanced higher. Even the addition of univalent counterions gradually causes a decrease in the electrical energy.

8.5 SOLUBILITY OF WATER-SOLUBLE POLYMERS

Biocolloids, polysoaps, partially saponified polyvinylacetate (PVA-Ac), and so on, consist of hydrophilic groups and nonpolar lipophilic groups. These hydrophilic polymers are either completely soluble in water, or are completely insoluble but swell in a fairly large amount of water. Many factors come into play in the solution of these polymers.

Polymers must be in a liquid state when they are solvated, otherwise solvent can not dissolve into polymer phase. The solubility of nonionic polymers such as partially saponified polyvinylacetate changes markedly with temperature, because the dipole interaction between the hydrophilic groups of polymer and water decreases with temperature rise. Hence, the temperature change in solutions of nonionic polymers corresponds to the change in the chemical nature

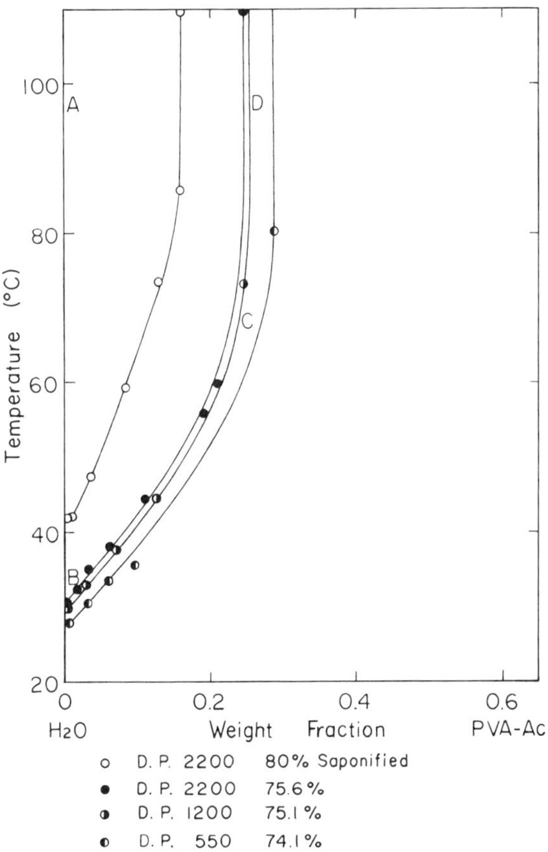

Fig. 8.8 Phase diagrams of water versus partially saponified poly(vinyl acetate) of different saponification degree and degree of polymerization (DP), as a function of temperature. From Ref. 15. Reprinted with permission. Copyright by the American Chemical Society.

of polymers. The solubility of higher temperature corresponds to the solubility of less hydrophilic polymer at medium temperature, and vice versa. The change of solubility of PVA-Ac in water will be described first, because the data are so suggestive.

It is considered that partially saponified polyvinyl acetate (PVA-Ac) resembles the block copolymers of vinylacetate and vinyl alcohol. The phase diagram of PVA-Ac-H_2O as a function of temperature is shown in Fig. 8.8 [15]. The curve ABCD represents the mutual solubility of water and PVA-Ac. It is clear from Fig. 8.8 that the liquid-liquid solubility curve is affected sensitively by the saponification degree, but not by the polymerization degree. Although the solution is so viscous and difficult to dissolve above 15-25 wt% (freeze-dried PVA-Ac is easier to dissolve), it is soluble in water over all compositions below the cloud point, B-C. The solution becomes cloudy and splits into two phases at temperatures above the cloud point curve. The one is a polymer phase containing a large amount of water, and the other is the water phase containing practically no polymer. The activity of water in this phase is close to unity.

The solubility of polymer in water phase is conventionally given by

$$\ln a_2 = \ln \phi_2 + \phi_1\left(1 - \frac{V_2}{V_1}\right) + \frac{V_2\phi_1{}^2 B'}{RT} \tag{8.25}$$

On the other hand, the solubility of water in the polymer phase is conventionally represented,

$$\ln a_1 = \ln \phi_1 + \phi_2\left(1 - \frac{V_1}{V_2}\right) + \frac{V_1\phi_2{}^2 B'}{RT} \tag{8.26}$$

where a is the relative activity, V_i are the molecular volumes of respective components in solution, ϕ_i are the volume fractions, and B' is the enthalpy of solution per unit volume [3]. It is expected from Eq. (8.25) that the solubility of polymer may be very small, because the molecular volume V_2, and therefore the heat of solution of polymer, is very large. Inversely, the molal volume of solvent V_1 is so small that the solubility of water in polymer is large.

The solubility of water in the polymer phase is already large (99.99 mol% or approximately 76 wt%) at point C. The dissolution of water into polymer proceeds with temperature depression because of the increase of iceberg formation and of hydration surrounding the solute molecule. Finally water and polymer mix with each other

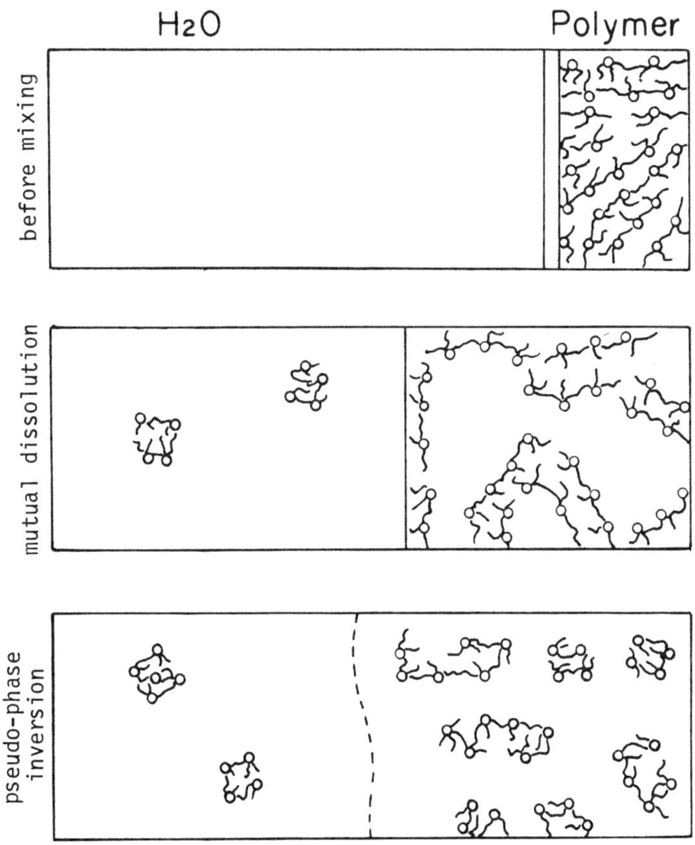

Fig. 8.9 Schematic representation of the dissolution of a water-soluble polymer in water. From Ref. 15. Reprinted with permission. Copyright by the American Chemical Society.

completely below the temperature B. The schematic diagram of the dissolution process in water of a polymer which consists of hydrophilic and lipophilic groups is shown in Fig. 8.9 [15,16].

It is considered that water dissolves infinitely into the polymer phase and that pseudo-phase inversion occurs in the polymer phase. If this happens, water becomes the continuous phase. Polymer molecules will orient, rearrange so as to decrease the free energy of mixing. This phenomenon is akin to the micellar dispersion of nonionic surfactant due to the increase of water solubility into the surfactant phase at low temperature. If the saponification degree of PVA-Ac is smaller, i.e., less hydrophilic, the BC curve will shift to lower temperature and the CD curve will shift to higher concentration, since the hydration of water per unit weight of polymer will decrease. Only swelling of water in polymer may occur below a certain saponification degree (approximately 70%). Although molecules are aggregated to micelles by physical bonds in the case of surfactant, and monomers are linked by covalent bond in polymers, the mechanism of dissolution in water is similar.

The mutual solubility of liquid in ordinary solution changes gradually with temperature. On the contrary, polymer (PVA-Ac) is practically insoluble in water above the cloud point, but it mixes with water over all compositions slightly below the cloud point. It may be concluded that the sudden change from complete solubility to insolubility with a small change in temperature is characteristic of polymers. Similar phenomena are observed, arising from the minute change in hydrophile-lipophile balance of biocolloids, such as proteins. For example, hydrophilic proteins may be completely soluble, but more lipophilic proteins may be completely insoluble, just as the nonionic polymer below and above the cloud point. The similar drastic change of solubility behavior is observed by the use of additives in solution. The temperature change plays a predominant role for the solubility behavior of nonionic polymers, whereas the change in chemical nature, i.e., hydrophile-lipophile balance of the molecule is the most decisive factor controlling solubility of both ionic and nonionic polymers in water.

REFERENCES

1. P. J. Flory, *Principles of Polymer Chemistry,* Cornell University Press, Ithaca, N. Y., 1953.
2. G. Gee and L. R. G. Treloar, Trans. Faraday Soc., *38*, 147 (1942); G. Gee and W. J. C. Orr, Trans. Faraday Soc., *42*, 507 (1946).
3. P. J. Flory, J. Chem. Phys., *9*, 660 (1941); *10*, 51 (1942).

4. M. L. Huggins, J. Chem. Phys., *9*, 440 (1941).
5. A. R. Miller, Proc. Camb. Phil. Soc., *38*, 109 (1942); *39*, 54 (1943); *39*, 131 (1939).
6. E. A. Guggenheim, Proc. Roy. Soc. A, *183*, 203 (1944).
7. J. H. Hildebrand, J. Chem. Phys., *15*, 225 (1947).
8. M. L. Huggins, J. Phys. Chem., *75*, 1255 (1971).
9. A. R. Shultz and P. J. Flory, J. Am. Chem. Soc., *74*, 4760 (1952).
10. G. Gee, "Thermodynamics of Rubber Solutions and Gels," in *Advances in Colloid Science, II*, Interscience, New York, 1946.
11. R. L. Scott, Thesis, Princeton University, 1945.
12. H. Mark and A. V. Tobolsky, *Physical Chemistry of High Polymeric Systems*, Interscience, New York, 1950, p. 263.
13. H. Burell, Interchem. Rev., *14*, 3 (1955).
14. S. A. Rice and M. Nagasawa, *Polyelectrolyte Solutions*, Academic Press, New York, 1961.
15. H. Kunieda and K. Shinoda, in *Colloidal Dispersions and Micellar Behavior*, ACS Symposium Series No. 9, edited by K. L. Mittal (1975), p. 278.
16. K. Shinoda, Bull. Chem. Soc. Japan, *35*, 1213 (1962).

EXERCISES

8.1 The vapor pressure of solvent in a polymer solution is expressed as follows

$$\frac{\Delta\bar{G}_1}{RT} = \ln \frac{P_1}{P_1^\circ} = \ln \phi_1 + \phi_2\left(1 - \frac{V_1}{V_2}\right) + \frac{BV_1\phi_2^2}{RT}$$

in which ϕ_1 and ϕ_2 are the volume fractions, and V_1 and V_2 are the molal volumes of the solvent and solute, respectively. B is an interaction parameter (of the heat of solution). Prove the Raoult's law may hold in dilute solution.

Answer: Since $\phi_1 \approx 1$, $\ln \phi_1$ may be expanded, and

$$\ln \frac{P_1}{P_1^\circ} = -\phi_2 - \frac{\phi_2^2}{2} - \frac{\phi_2^3}{3} + \phi_2\left(1 - \frac{V_1}{V_2}\right) + \frac{BV_1\phi_2^2}{RT} + \cdots$$

$$= -\frac{V_1}{V_2}\phi_2 + \left(\frac{BV_1}{RT} - \frac{1}{2}\right)\phi_2^2 - \frac{\phi_2^3}{3} + \cdots$$

In a sufficiently dilute solution, terms above the second order may be ignored, and since $x_1V_1 >> x_2V_2$,

$$\ln \frac{P_1}{P_1^\circ} \approx -x_2 \approx \ln(1 - x_2) = \ln x_1$$

which means

$$P_1 = P_1^\circ x_1$$

PROBLEMS

8.1 Suppose that the entropy of mixing of a solution containing n_1 moles of a liquid of molal volume V_1 and n_2 moles of a liquid with molal volume V_2 is given by

$$\Delta S_M = -R\left(n_1 \ln \frac{n_1V_1}{n_1V_1 + n_2V_2} + n_2 \ln \frac{n_2V_2}{n_1V_1 + n_2V_2}\right)$$

Derive the expression for the partial molal entropy of the second component.

8.2 Derive the critical temperature and critical composition from the equation which expresses the change of activity with composition,

$$RT \ln a_2 = RT\left[\ln \phi_2 + \phi_1\left(1 - \frac{V_2}{V_1}\right)\right] + V_2\phi_1^2(\delta_1 - \delta_2)^2$$

The experimental critical composition, which depends on the shape and flexibility of the molecule, will be a value intermediate between this equation and that given by Eq. (5.16).

chapter nine

SURFACTANT SOLUTIONS

9.1 CHARACTERISTIC PROPERTIES OF SURFACTANT SOLUTIONS

Molecules with long hydrocarbon chains are generally insoluble in water. However, in contradiction to this common-sense observation, surfactants (which contain paraffin chains in their molecules), dissolve in water over almost all compositions. This is a remarkable phenomenon in the field of solution, and suggests that, for these compounds, there is an entirely different mechanism of dissolution.

"Surface-active agent" or "surfactant" are generic terms for a group of substances whose molecules contain a hydrophilic group at one end and a lipophilic group at the other. Thousands of variations on this simple arrangement are possible, depending on the sizes, number, types, and combinations of the hydrophilic and lipophilic groups including connecting groups. They are classified, according to their ionic nature, as anionic, cationic, ampholytic, zwitterionic, and nonionic surfactants. Some are oil-soluble, but not water-soluble, and vice versa. Some are biological substances, such as cholates, taurocholates, monoglycerides, lecithin, and so on.

Since the surfactant molecule is composed of hydrophilic and lipophilic groups, it has several properties characteristic of this combination, such as surface activity at the water-oil or water-air interface, orientation of adsorbed molecules, micelle formation due

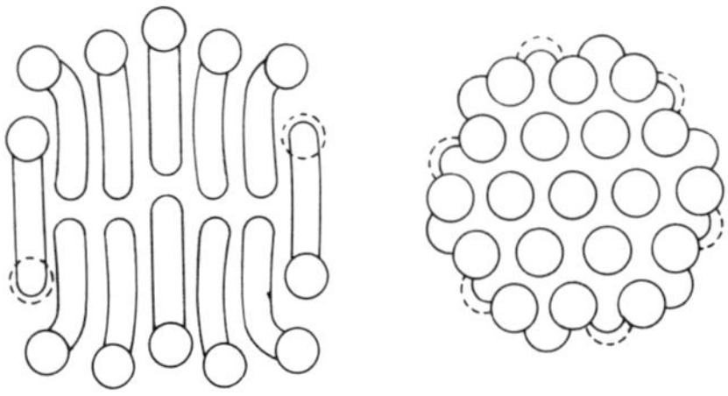

Fig. 9.1 Surfactant micelle. The micelle can be thought of as either spherical, spheroidal, or laminar, depending on the aggregation number.

to the aggregation of scores of molecules above the critical concentration for micelle formation (cmc), solubilization of third substances, and formation of liquid crystals.

In the micelle the hydrophilic groups are oriented outwards and the lipophilic groups are oriented inwards in aqueous solution, while the reverse situation occurs in hydrocarbon media. A model of a micelle is shown in Fig. 9.1.

The interior of the micelle resembles a liquid paraffin. When an oil or an oil-soluble substance is added to a micellar solution, it is solubilized into the micelle. This phenomenon corresponds to partition equilibrium between an aqueous phase and a micellar phase. Polar oils, such as alcohols and amines, penetrate into the palisade layer of the surfactant. Some kinds of dyes may be adsorbed on the micelle surface.

It is significant that surfactants form micelles, i.e., molecules orient, aggregate, and form structures in solution. This phenomenon is entirely different from ordinary solution in a state of random mixing. As a result of the formation of these structures, unique new properties appear, such as solubilization, mentioned above. Biological systems also possess structure, and several different molecules cooperate and exhibit new specific properties. Surfactant solutions, in this respect, mediate between ordinary solutions and biological systems.

9.2 PSEUDO-PHASE SEPARATION MODEL FOR MICELLE FORMATION [1,2]

The plot of surface tension versus logarithm of the concentration of the pure nonionic surfactants, α-d-octylglucoside and α-d-decylglucoside are shown in Fig. 9.2. Initially, the surface tension of the solution decreases almost linearly with the logarithm of the concentration, but then remains nearly constant, i.e., the surface tension scarcely decreases further with concentration. If we apply the Gibbs adsorption isotherm

$$\left(\frac{\partial\gamma}{\partial \ln a_2}\right)_T = \left(\frac{\partial\gamma}{\partial \ln C_2}\right)_T\left(\frac{\partial \ln C_2}{\partial \ln a_2}\right)_T = -RT\Gamma_2 \tag{9.1}$$

to the data, the surface excess is almost constant below the cmc, but becomes almost zero above, provided $(\partial \ln C_2/\partial \ln a_2)_T = 1$.

In dilute solution, the activity is usually proportional to the concentration, and $(\partial \ln C_2/\partial \ln a_2) = 1$ holds fairly accurately.

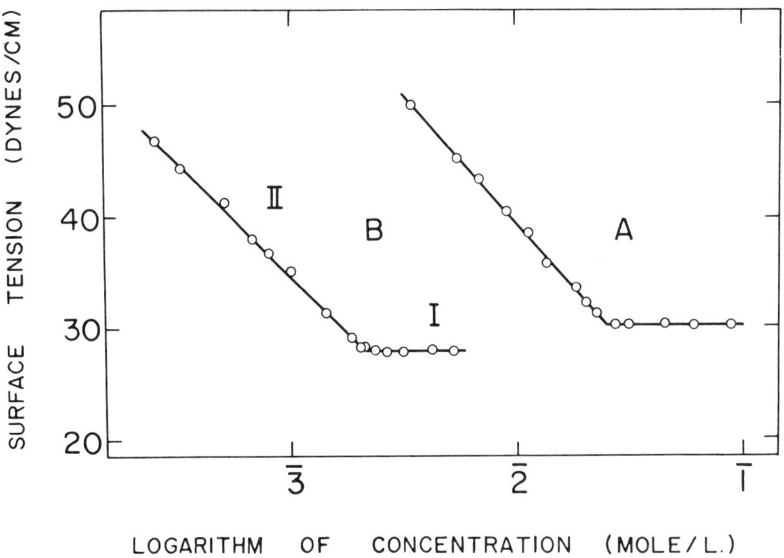

Fig. 9.2 Surface tension versus concentration for octylglucoside (A) and decylglucoside (B). Reprinted with permission from Bull. Chem. Soc. Japan, *34*, 273 (1961).

However, it is evident from various phenomena, as well as from the results of radio-tracer studies, that the surface excess of surfactants remains constant above the cmc.

If Γ_2 is finite, and $\partial\gamma/\partial \ln C_2 \approx 0$, then $\partial \ln a_2/\partial \ln C_2 \approx 0$. This means that the activity stays nearly constant when the concentration of surfactant is changed.

In ordinary dilute solutions which obey Henry's law, this is not true. However, in the case of sparingly soluble substances such as C_6H_6 or $C_{10}H_{21}OH$ in water, the activity may increase almost in proportion to the concentration of solute up to the saturation concentration of singly dispersed species, but it does not increase after the excess solute phase appears. From the surface tension versus concentration data, we are forced to conclude that micelle formation resembles the phase separation phenomenon, and that the cmc is the saturation concentration of the molecularly dispersed species. Measurements of osmotic coefficients are consistent with this model.

The important distinction between micelle formation and true phase separation is that the micellization does not lead to an infinite aggregation number. Thus, the micellar solution is an optically transparent single phase, but its thermodynamic properties resemble those of a two-phase system. The partial molal enthalpy, entropy, free energy, etc., remain almost constant with concentration above the cmc, supporting the pseudo-phase dispersion model. Below the cmc, the solvent and solute obey the laws of dilute solution.

9.3 SOLUBILITY OF IONIC SURFACTANTS IN WATER [1]

When the concentration of an ionic surfactant in aqueous solution is increased up to the solubility limit, hydrated solid surfactant separates below the melting point of the hydrated surfactant (Krafft point). However, the solubility rapidly increases with temperature, and at a temperature approximately 4°C above the Krafft point, a solution of almost any composition becomes a single homogeneous phase.

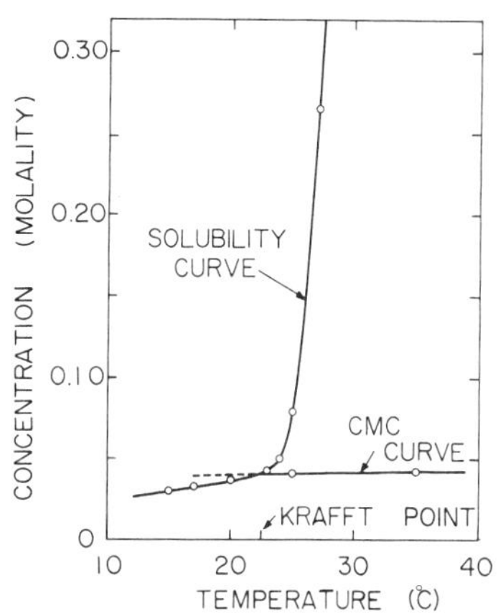

Fig. 9.3 The relation of the solubility curve, cmc curve, and Krafft point for sodium decylsulfonate.

When the concentration of surfactant reaches the saturation concentration of a molecular (or ionic) dispersion above the Krafft point, scores of molecules (or ions) associate to form a particle called a micelle. The micelle particles are in a liquid state, and reach equilibrium at a certain association number. They do not associate infinitely, the size of the micelle is significantly smaller than the wavelength of visible light, and the solution appears transparent. It can be regarded as a system in which the particles corresponding to the pseudo-phases are dispersed in the solution. In other words, the hydrated solid surfactant melts into liquid when it reaches the melting point, and disperses as a micelle. Consequently, the solubility increases very rapidly as shown in Fig. 9.3.

It may be convenient to consider Fig. 9.3 by analogy with the phase equilibria of benzene or dodecanol in water. The micelle corresponds to liquid benzene, the solid surfactant to solid benzene, and so on. As the activity of surfactant changes but little with concentration, since micelle formation resembles phase separation, the solubility increases rapidly. It is therefore clear that the Krafft point is, in fact, the melting point of the hydrated solid surfactant. In fact, the value of the Krafft point rises with chain length in the same way as the melting points of a series of organic homologues. When the counterion is varied, the Krafft point generally becomes higher as the ionic nature (bond) decreases. The Krafft point is depressed by the addition of, for example, alcohols and other surfactants, just as are the melting points of ordinary substances. The Krafft point is constant at a constant composition, temperature, and pressure (concentration), as required by the phase rule. One of the important features of Fig. 9.3 is the fact that the equilibrium curve between solid-solute and solid-liquid (i.e., micelle) represents the solubility curve, whereas in the case of ordinary systems, such as water-benzene, the curve between solid-solute and liquid-solute is the solubility curve. Thus, the physical meaning of the solubility curve is different.

This totally new interpretation has been confirmed by the measurement of the partial molal volume of surfactant above and below the Krafft point, the heat of solution above the Krafft point, and of the heat of hydration. The reason that the solubility does not increase perfectly vertically beyond the Krafft point is that the Krafft point rises slightly with increase in counterion concentration, and that the association number of the micelle is not infinite. This mechanism of dissolution is characteristic of associated colloidal solutions.

It may be assumed that complete mixing of molecules occurs below the saturation concentration of molecular dispersion and that the micellar molecules can be treated as if they were perfectly oriented. Thus, the solubility below the Krafft point and the critical micelle concentration can be explained by the concept of the regular solution. Also, our intuitive prediction that molecules containing a long hydrocarbon chain would not readily dissolve in water is corroborated by the very low critical micelle concentration, i.e., solute-liquid equilibrium.

9.4 SOLUBILITY OF NONIONIC SURFACTANTS IN WATER [3]

An aqueous solution of a nonionic surfactant becomes suddenly cloudy above a certain temperature, called the *cloud point*, and the solution splits into two phases. One phase is a water phase, which contains a very small amount of singly dispersed nonionic surfactant (approximately 0.02 wt%); the other is a surfactant phase containing dissolved water. Thus, a cloud-point curve is a liquid-liquid solubility curve, with a lower consolute temperature. The cloud point curve of octaoxyethylene dodecyl ether in water is shown in Fig. 9.4.

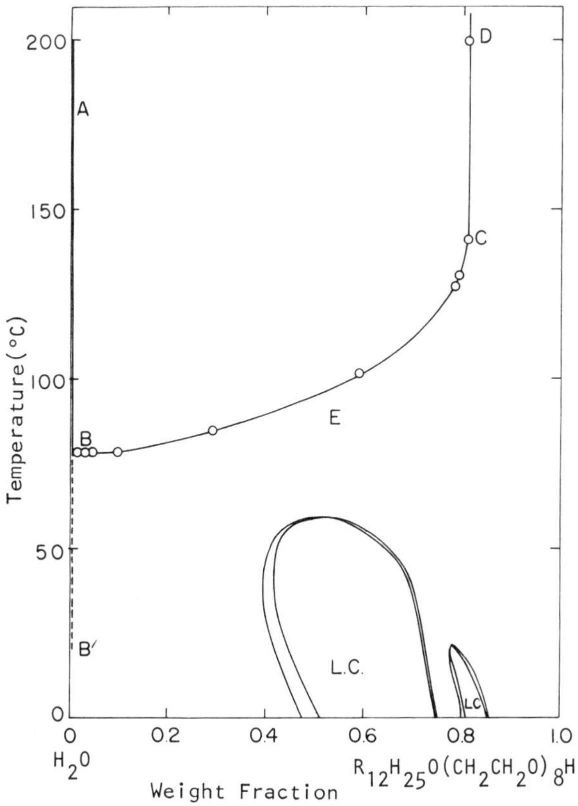

Fig. 9.4 Phase diagram for the system H_2O-$C_{12}H_{25}O(CH_2CH_2O)_8H$. From Ref. 3. Reprinted with permission.

Part A-B of the solubility curve is extremely close to the water axis, indicating the very small solubility of paraffin-chain compounds in water. The solubility at this temperature is equal to the saturation concentration of singly dispersed molecules. The critical micelle concentration below the cloud point corresponds to this saturation concentration. The cmc value of $C_{12}H_{25}O(CH_2CH_2O)_6H$ is, for example, as low as 8.7×10^{-5} mol/liter at 25°C [4]. Although the saturation concentration of a singly dispersed nonionic surfactant in water gradually increases with temperature depression {for example, from 6.4×10^{-4} mol/liter at 45°C to 11.4×10^{-4} mol/liter at 15°C in the case of $C_{10}H_{21}O(CH_2CH_2O)_6H$ [4]}, the absolute value is still very small. In other words, the solubility-temperature curve of the nonionic surfactant in water below the cloud point is practically unchanging, as shown by the dotted curve BB' in Fig. 9.4. The physical meaning of the curve AB is the same as that of an ordinary solubility curve.

On the other hand, the solubility of water in the nonionic surfactant phase is fairly large, i.e., about 19 wt% (87 mol%) at 140°C because the molal volume of water is so small that it dissolves with far less energy per molecule. The solubility of water in surfactant does not vary much over the temperature range between C and D, as shown in Fig. 9.4. The increased thermal motion is probably cancelled by the decreased dipole-dipole interaction. The physical significance of the CD portion of the liquid-liquid solubility curve is similar to that of the solubility curve of solvent in polymer. As the solubility of water in octaoxyethylene dodecyl ether is large (about 87 mol% at point C), it may increase to infinity at the lower temperature at which the iceberg formation of water and the hydration between the water and the ether-oxygens of the hydrophilic group increases.

Let us consider the dissolution process for a surfactant below its cloud point. Water at first solvates the hydroxy group and the ether-oxygen atoms of the hydrophilic moiety of the surfactant molecule. If more water is dissolved, the surfactant phase will have a layered structure, as in the neat phase, and water and surfactant are both continuous in this composition range. If more water is dissolved, the surfactant phase may disperse as bimolecular leaflets (micelles). If pseudo-phase inversion of the biomolecular leaflet occurs close to the point E, surfactant phase will mix with water phase, the composition of which is expressed by the curve BB'. In other words, the solution between B and E is to be regarded as a mixture of the water phase and the hydrated surfactant aggregate. The physical meaning of curve BE is that above this curve (temperature), the surfactant aggregates infinitely, but below the curve the surfactant aggregates in finite structures, and disperses as a stable colloid. This may be the reason that the curve CEB intersects curve ABB' perpendicularly and discontinuously. If the physical significance of the two curves was identical, they would be continuous. We can conclude that the curve ABECD is a liquid-liquid solubility curve in that the physical significance changes discontinuously at point B and around E.

The most distinct feature of the vapor pressure versus composition data of the water-hexaoxyethylene dodecyl ether system is that Raoult's law does not hold for very dilute solutions, and the vapor pressure remains nearly constant. The water activity is shown as a function of weight fraction in Fig. 9.5 [5].

Applying the Gibbs-Duhem relation,

$$x_1 \frac{\partial \ln f_1}{\partial x_1} - x_2 \frac{\partial \ln f_2}{\partial x_2} = 0 \qquad (9.2)$$

we conclude that the activity of the surfactant hardly changes with concentration up to about 40 wt%. This important conclusion, which also follows from the surface tension-concentration relation, supports the pseudo-phase separation model (two-phase model).

Micellar dispersion below the cloud point curve and phase separation above the curve can then be explained as follows: (1) the solubility of water in the surfactant phase increases with decrease in temperature, and finally becomes infinite below the cloud point,

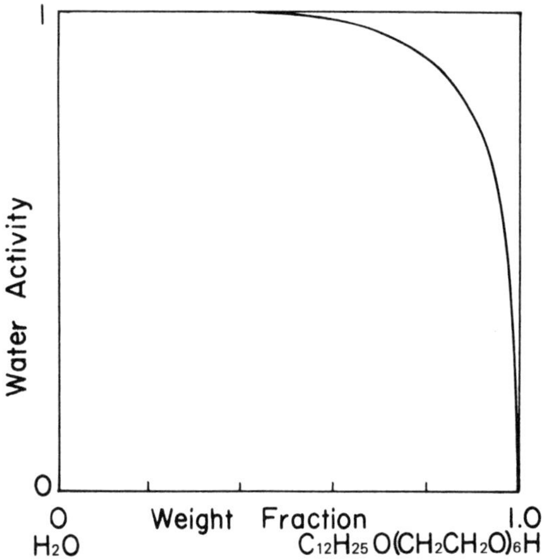

Fig. 9.5 Water activity versus weight fraction of $C_{12}H_{25}O(CH_2CH_2O)_6H$ at 25°C

i.e., micellar dispersion is considered to be due to the increase of water solubility in the surfactant phase; and/or (2) the curvature of the surfactant monolayer toward the water changes from concave to convex at the cloud point owing to the lowering of the temperature, and thus a surfactant phase which separates above the cloud point, disperses as micelles below.

9.5 CRITICAL MICELLE CONCENTRATION OF IONIC SURFACTANTS [1]

Since the cmc is the saturation concentration of molecular (ionic) dispersion in equilibrium with pseudo-phase, it can be treated by applying the concepts of regular solution. The following is an example of the application of regular solution theory to a nonregular solution.

We start with the basic solubility equation, (4.10),

$$RT \ln a_2 = RT \ln x_2 + V_2\phi_1^2 B \tag{4.10}$$

The micelle is in a liquid state, and the solubility of water in the micelle may be quite small so that the activity of the micellar surfactant relative to a pure liquid state may be assumed to be unity. This means the left-hand side of Eq. (4.10) is equal to zero. Hence, the free energy term due to dilution, RT ln x_2, will balance the free energy increase due to mixing (enthalpy) term. Thus, the cmc may be calculated by estimating the free energy change due to the dissolution of surfactant molecule (or ion) from the micelle into the bulk solution.

Since the system being discussed is nonregular, the energy of mixing should be determined experimentally. However, direct calorimetric measurements of the enthalpy of mixing are not really useful, because the enthalpy change is frequently obscured by accompanying abnormalities. In order to calculate the cmc, one must know the free energy change due to the transfer from micelle to solution.

The free energy change due to the association of the hydrocarbon chains and the electrical energy involved in the aggregation of charged groups are the two principal factors in the process of micelle formation. If the nonelectrical energy decrease attributable to the association per methylene group is expressed by ω, that per molecule is expressed by $m\omega$ (m is the number of chain carbons), since the energy decrease is proportional to the chain length. If the surface potential of the micelle surface is ψ_0, an electrical energy $e\psi_0$ per micelle ion is required. On the other hand, gegenions are also attracted and approach the micelle surface, so that the total electrical energy required to form the micelle is lessened. This effect is corrected for by a factor $1 > K_g > 0$, which multiplies the surface energy. Substituting these values (and assuming $a_2 = 1$) into Eq. (4.10), we obtain

$$\ln a_2 \approx 0 = \ln x_2 + \frac{m\omega - K_g e\psi_0}{kT} \phi_1^{\ 2} \tag{9.3}$$

Since cmc = x_2 is very small, ϕ is close to 1.

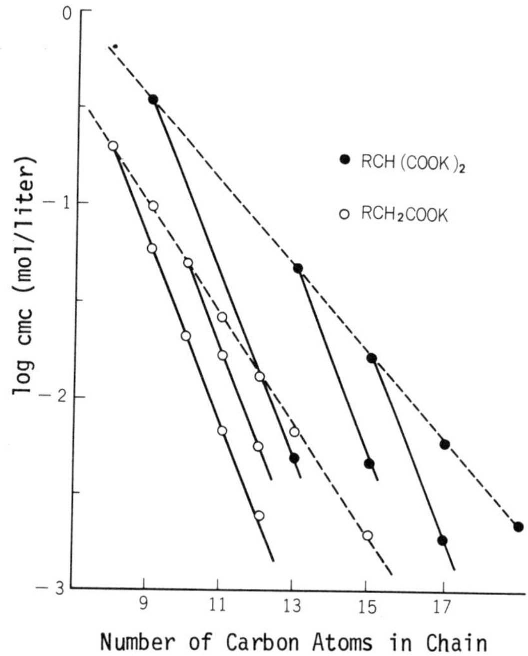

Fig. 9.6 The relation between the log cmc and the number of carbon atoms in the hydrocarbon chain at a fixed concentration of gegenions.

From the Poisson-Boltzmann equation, the electrical energy per micelle-forming ion is estimated to be [1]

$$K_g e\psi_0 = E_{el} = K_g kT\left(\ln \frac{36\pi\sigma^2}{DNkTx_3}\right) = K_g kT\left(\ln \frac{2000\pi\sigma^2}{DNkTC_3}\right) \qquad (9.4)$$

where σ is the surface charge density, D is the dielectric constant, and x_3 and C_3 are the mole fraction and equivalent concentration of gegenions, respectively. This energy is constant for a homologous series of surfactants (σ is constant) when the gegenion concentration is constant. As shown in Fig. 9.6, the cmc is depressed by about one-third by the addition of one methylene group at constant gegenion concentration. From this observation, we may deduce that the free energy change involved in transferring hydrocarbon from the hydrocarbon environment to the aqueous phase is ω = kT ln 3 per methylene group.

Substituting this relation, as well as Eq. (9.4), into Eq. (9.3), we obtain

$$\ln x_2 = -m \ln 3 + K_g \ln \frac{36\pi\sigma^2}{DNkTx_3} \tag{9.5}$$

Expressing the concentration as molality (in dilute solution),

$$\ln C_2 = -m \ln 3 - K_g \ln C_3 + K_1 \tag{9.6}$$

On the assumption $\omega = kT \ln 3$, Eq. (9.6) describes the change of cmc with hydrocarbon chain length of the surfactant. It also shows that the cmc is inversely proportional to the K_gth power ($K_g \approx 0.6$) of the gegenion concentration. When no salt is added, $C_2 = C_3$, and we obtain

$$\ln C_2 = -\frac{m}{1 + K_g} \ln 3 + K_0$$

$$= -\frac{\omega}{kT}\frac{m}{1 + K_g} + K_0 \tag{9.7}$$

The constant K_g is found to be in the range 0.5 to 0.65, independent of the nature of the polar groups. Hence, the cmc decreases by almost one-half per one methylene group increase in the absence of added salts. The effect of the gegenion is, roughly speaking, independent of its nature, as long as the valence is the same. Cmc values for typical ionic surfactants are summarized in Table 9.1.

More complete treatments of theory and comparison with experimental data on the effect of molecular structure, e.g., presence of two hydrophilic groups, two hydrocarbon chains, change in position of polar group(s), effect of gegenion valency, mixing of surfactants and of additives on the cmc, are discussed in the literature [1].

For example, when the gegenion is divalent, as in the case of $R_nSO_4 \cdot 1/2$ Mg, the electric potential on the micelle surface is reduced by one-half:

$$K_g e\psi_0 = E_{el} = \frac{K_g kT}{Z_i}\left(\ln \frac{2000\pi\sigma^2}{DNkTC_i}\right) \tag{9.8}$$

Table 9.1 Critical Micelle Concentrations of Ionic Surfactants in mol/liter at 25°C [1]

Surfactants	cmc	Surfactants	cmc
$R_{12}SO_4Na$	0.008	$R_8CH(COOK)_2$	0.35
$R_{12}SO_3Na$	0.010	$R_{10}CH(COOK)_2$	0.13
$R_{12}C_6H_4SO_3Na$	0.0012	$R_{12}CH(COOK)_2$	0.048
R_7COOK	0.40	$R_{14}CH(COOK)_2$	0.017
R_9COOK	0.10	$R_{16}CH(COOK)_2$	0.0063
$R_{11}COOK$	0.026	$R_{12}NH_3Cl$	0.014
$R_{13}COOK$	0.0066	$R_{12}N(CH_3)_3Cl$	0.016
$R_{15}COOK$	(0.0016)	$C_5F_{11}COOK$	0.5
$R_{12}SO_4Li$	0.009	$C_7F_{15}COOK$	0.027
$R_{12}SO_4K$	0.008	$(C_9F_{19}COOK)$	(0.0009)
$R_{12}SO_4Cs$	0.007	$H(CF_2)_8COONH_4$	0.038
$R_{12}SO_4N(CH_3)_4$	0.0055	$H(CF_2)_{10}COONH_4$	0.009

where Z_i is the valency of the gegenion. Substituting Eq. (9.8) into Eq. (9.3), we obtain

$$\ln C_2 = -m \ln 3 + \frac{K_g}{Z_i}\left(\ln \frac{2000\pi\sigma^2}{DNkTC_i}\right) + K_2 \tag{9.9}$$

In the absence of added salt ($C_2 = Z_iC_3$), the relation between the cmc of the divalent salt and the hydrocarbon chain length is

$$\ln C_2 = \frac{1}{1 + K_g/Z_i}\left(-m \ln 3 + \frac{K_g}{Z_i} \ln \frac{2000\pi\sigma^2 Z_i}{DNkT}\right) + K_3 \tag{9.10}$$

The conclusion that the slope of the plot of ln cmc vs. m of a 1:2 type of ionic surfactant lies between that of a 1:1 and a nonionic surfactant is borne out by Fig. 9.7.

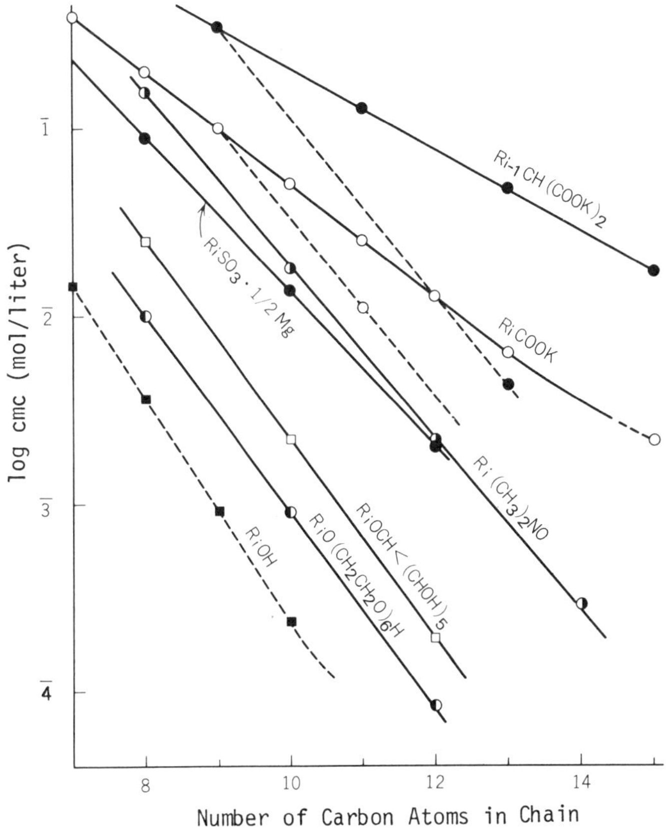

Fig. 9.7 Comparison of cmc values of various types of surfactants as a function of the number of carbon atoms in the hydrocarbon chain.

9.6 CRITICAL MICELLE CONCENTRATION OF NONIONIC SURFACTANTS

The saturation concentration of ionic surfactants in equilibrium with micelles is comparatively large, because the nonelectrical energy decrease due to the aggregation is diminished by the electrical work. In the case of nonionic surfactants, there is no such electrical repulsion between the molecules, so that the saturation concentration of molecular dispersion, i.e., the cmc, is much smaller than that of an ionic surfactant.

Let us consider the cmc of nonionics, taking a polyoxyethylene alkyl aryl ether type as an example. Hydration takes place at the hydrophilic moiety, and it may not differ much in the bulk of the solution or in the aggregated state. Namely, the energy required to gather the hydrophilic groups into the outer core of the micelle seems to be small. As shown in Tables 9.2 and 9.3, the effect of the size of the hydrophilic group on the cmc is small, but the hydrocarbon chain length has a marked effect.

As no electrical energy is needed in the micellization of a nonionic surfactant, we obtain

$$\ln x_3 = -\frac{m\omega}{kT} + K_4 = -m \ln 3 + K_4 \tag{9.11}$$

or

$$\ln C_2 = -m \ln 3 + K_4' \tag{9.12}$$

in place of Eqs. (9.6) and (9.7).

Since K_4' is not large, the cmc value is mainly dependent on the hydrocarbon chain length. It is evident from Eqs. (9.7) and (9.12) that the cmc values of nonionic surfactants are roughly the

Table 9.2 The Relation Between the Number of Oxyethylene Chain and the cmc of Nonionic Surfactants at 25°C

$i\text{-}C_9H_{19}C_6H_4O(CH_2CH_2O)_nH$ [6]		$i\text{-}C_8H_{17}C_6H_4O(CH_2CH_2O)_nH$ [7]	
n	cmc (10^{-4} mol/liter)	n	cmc (10^{-4} mol/liter)
9.5	0.8-0.9	3	1.0
15	1.1-1.3	4	1.3
20	1.4-1.7	5	1.7
30	1.8-3.0	6	2.5
50	2.8	7	2.7
100	10	8	2.8
		9	3.0
		10	3.2

Table 9.3 Relation Between Hydrocarbon Chain-length and cmc of Polyoxyethylene Alkylethers and Dimethylalkylamine Oxides [8-10]

Surfactants	cmc (10^{-4} mol/liter)	T (°C)
$R_6O(CH_2CH_2O)_4H$	900	20
$R_8O(CH_2CH_2O)_yH$	98	25
$R_{10}O(CH_2CH_2O)_5H$	8.6	20
$R_{12}O(CH_2CH_2O)_6H$	0.82	20
$R_{16}O(CH_2CH_2O)_6H$	0.01	25
$R_{16}O(CH_2CH_2O)_6H$	0.99	27
$R_{10}(CH_3)_2NO$	160	27
$R_{12}(CH_3)_2NO$	21	27
$R_{14}(CH_3)_2NO$	2.8	27

$(1 + K_g)$th power of those of comparable ionic surfactants, provided that the hydrocarbon chain length of both surfactants are the same.

In the case of nonionic surfactants, the slope of the curve relating logarithm of the cmc and the number of carbon atoms in the chain m is equal to $-\omega/kT$, whereas in ionic surfactants with one or two dissociable groups, the slopes are $-\omega/(1 + K_g)kT$ and $-\omega/(1 + 2K_g)kT$, respectively. In the case of ionic surfactants of divalent salts, the slope is $-\omega/(1 + K_g/2)kT$, as shown in Fig. 9.7. All these relations follow theoretically from Eqs. (9.6) to (9.12).

9.7 ENTHALPY AND ENTROPY OF MICELLIZATION

Let us denote water as component 1, surface-active ion as component 2, gegenion as component 3, and simili-ion (i.e., non-surface-active ion of same charge as surface-active one) as component 4. The electrochemical potential of surface-active ions in the micellar state is equal to those in the bulk solution,

$$\bar{\mu}_2^M = \bar{\mu}_2^\ell \tag{9.13}$$

Since the electrochemical potential of surface-active ions in the micellar state will change only slightly with the concentration of simili-ions, the potential is a function only of T, x_2, x_3 under conditions of constant pressure. If we denote the change of the electrochemical potential of surface-active (micelle-forming) ions on micellization as $\Delta\bar{G}_m$, the total differential of $\Delta\bar{G}_m$ is

$$d(\Delta\bar{G}_m) = \left(\frac{\partial\Delta\bar{G}_m}{\partial T}\right)_{p,x_2,x_3} dT + \left(\frac{\partial\Delta\bar{G}_m}{\partial \ln x_2}\right)_{p,T,x_3} d \ln x_2 + \left(\frac{\partial\Delta\bar{G}_m}{\partial \ln x_3}\right)_{p,T,x_2} d \ln x_3 \qquad (9.14)$$

It is evident from Eq. (9.13) that $\Delta\bar{G}_m = 0$.

$$\Delta S_m = \frac{\Delta H_m}{T} \qquad (9.15)$$

On the other hand,

$$\Delta S_m = -\left(\frac{\partial\Delta\bar{G}_m}{\partial T}\right)_{p,x_2,x_3} \qquad (9.16)$$

From Eqs. (9.14) to (9.16) we obtain

$$\Delta S_m = \frac{\Delta H_m}{T} = \left(\frac{\partial\Delta\bar{G}_m}{\partial \ln x_2}\right)_{p,T,x_3}\left(\frac{\partial \ln x_2}{\partial T}\right)_{p,\Delta\bar{G}_m} + \left(\frac{\partial\Delta\bar{G}_m}{\partial \ln x_3}\right)_{p,T,x_2}\left(\frac{\partial \ln x_3}{\partial T}\right)_{p,\Delta\bar{G}_m} \qquad (9.17)$$

From Eqs. (9.3) and (9.4),

$$-RT \ln a_2 = \Delta\bar{G}_m = -RT \ln x_2 - K_g RT \ln x_3 + \text{constant } (P, T) \qquad (9.18)$$

where K_g, it will be remembered, is a constant which can be obtained from the slope of the log cmc-log x_3 curve. Substituting Eq. (9.18) into (9.17), we obtain

$$-\Delta S_m = -\frac{\Delta H_m}{T} = RT\left(\frac{\partial \ln x_2}{\partial T}\right)_{p,\Delta\bar{G}_m} + K_g RT\left(\frac{\partial \ln x_2}{\partial T}\right)_{p,\Delta\bar{G}_m} \tag{9.19}$$

In the case where no salt is added, the number of surfactant ions is equal to the number of gegenions. Thus, substituting $x_3 = x_2$,

$$-\Delta S_m = -\frac{\Delta H_m}{T} = (1 + K_g)RT\left(\frac{\partial \ln x_2}{\partial T}\right)_{p,\Delta\bar{G}_m} \tag{9.20}$$

When x_3 is maintained constant by the addition of salt, the second term in Eq. (9.19) becomes equal to zero, and

$$-\Delta S_m = -\frac{\Delta H_m}{T} = RT\left(\frac{\partial \ln x_2}{\partial T}\right)_{p,x_3,\Delta\bar{G}_m} \tag{9.21}$$

The same value for ΔS_m is obtained experimentally from either Eq. (9.20) or (9.21), under the appropriate conditions. When partially dissociated aggregates are formed, the multiplicative constant $(1 + K_g)$ must be used as shown in Eq. (9.20). It has the value 1 in the case of nonelectrolytes and is equal to 2 for 1:1 electrolytes.

Critical micelle concentrations are usually very small. They range from 10^{-6} to 10^{-4} in mole fraction units. Although a large entropy of solution (18 to 27 eu) would be expected in a system of such low solubility, the entropy is only of the order of -2 to +4 eu for aqueous surfactant systems. Namely, the change of cmc with temperature is unexpectedly small. These results are ascribed to the iceberg formation of water molecules around the hydrocarbon chain [11]. If a large number of water molecules surrounding a solute molecule solidify, the enthalpy and entropy of solution will decrease considerably, and roughly equally, just as the enthalpy and entropy decrease in solidification [12]. Therefore, although the enthalpy and entropy of solution decrease substantially, the saturation concentration of molecular dispersion does not change very much, close to the temperature of iceberg formation. This phenomenon seems quite general when hydrocarbon chains (solute) dissolve in water, and the same phenomenon is observed with alcohols and amines. When water forms ice at atmospheric pressure, the number of nearest neighbors is four because of hydrogen-bond formation, and the water molecule occupies a larger volume than it does in the liquid state. However, in iceberg formation, the volume occupied by the water decreases, as it does in most ordinary substances, because the water molecules may pack more closely around the hydrocarbon chain.

9.8 SOLUTION OF IONIC SURFACTANTS IN NONAQUEOUS SOLVENTS [13]

In the present discussion, we limit nonaqueous solvents to hydrocarbons and their chlorinated derivatives. Liquids of high dielectric constant are excluded. Unlike aqueous solutions, the cohesive energy per unit volume is small for nonaqueous solvents. As a result, association occurs by the interaction between the hydrophilic groups of the surfactant. The hydrophilic group is oriented inward and the lipophilic outward in the micelle.

The principal driving force for micelle formation is the interaction among the hydrophilic groups. Since the lipophilic group and the solvent are similar, the type and length of the hydrophilic group does not have a big effect on the cmc, although there is a small increase in cmc with increasing chain length.

The cmc, the Krafft point, and the cloud point are the main factors controlling solubility in aqueous solution. However, in nonaqueous solutions of dinonylnaphthalene sulfonates [14], aggregates of low degree of association form at very low concentrations, almost independent of the concentration. The system may be thought of as a solution in which the aggregated particles dissolve as the solute, and the solubility of the micelles is governed by the interaction of the lipophilic portion with the solvent [14,15]. The liquid-liquid solubility curve of dinonylnaphthalene sulfonate and dimethylsiloxane heptamer is shown in Fig. 9.8.

There are cases in which the surfactant is barely soluble in the solvent, and the solvent rather dissolves in the surfactant

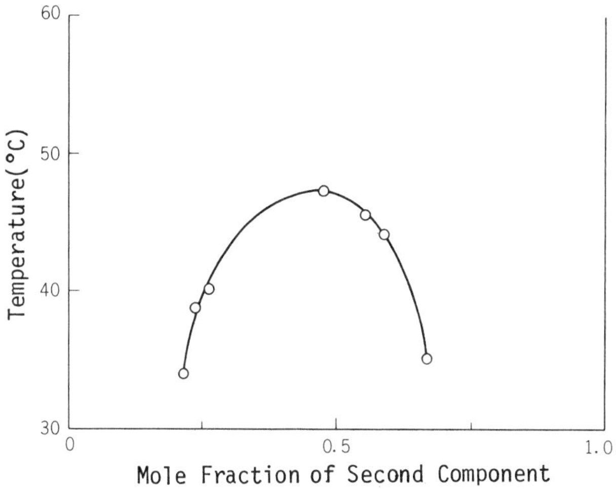

Fig. 9.8 The liquid-liquid solubility curve for dimethylsiloxane and lithium dinonylnaphthalene sulfonate.

phase, forming a gel or liquid crystal. In such a case, solubility can be increased by using a solvent whose molecular size and heat of solution in surfactant is small.

Since, as noted above, the main factor for the formation of aggregates in nonaqueous solvent is the interaction of the polar groups, the dissolution state is, in many cases, markedly affected by a small amount of added water. The addition of as little as 0.05% of water to a benzene solution of aryl stearate soaps converts the viscous fluid formed in the complete absence of water to a mobile liquid having a relative viscosity slightly larger than that predicted for the system by the Einstein equation for the viscosity of a suspension of spheres [16].

9.9 SOLUTIONS OF NONIONIC SURFACTANTS IN HYDROCARBONS

Since nonionic surfactant is generally in the liquid state in the presence of hydrocarbons, the solution may be treated as a liquid-liquid mixture. Examples of the mutual solubility of hydrocarbons and nonionic surfactants are shown in Fig. 9.9 [17].

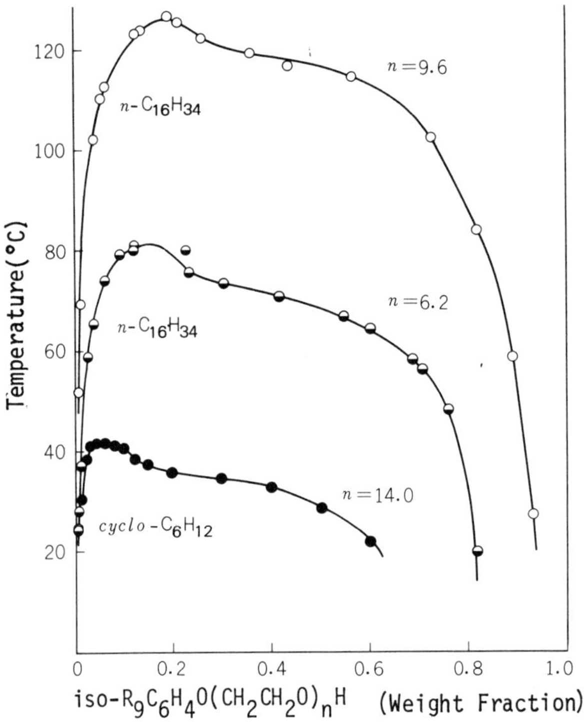

Fig. 9.9 The liquid-liquid solubility curves for polyoxyethylene nonyl ethers in n-hexadecane and in cyclohexane [17].

It is evident from the temperature dependence of the solubility in hydrocarbons that temperature has the opposite effect from that in water. The liquid-liquid solubility curves do not show a discontinuity, and this implies that the aggregation number is small, if, indeed, there is any aggregation at all. Addition of a small amount of water decreases the solubility of the surfactant, but promotes the formation of micelles [18]. The aggregation number may increase with temperature depression and particularly rapidly at a given temperature, owing to the enormous increase in solubilization of water. Under these conditions, the surfactant may be considered to be aggregating almost infinitely, and below this temperature it aggregates infinitely and phase separation takes place [19]. One phase is an oil phase and the other is a surfactant phase in which a large amount of oil and water is dissolved.

REFERENCES

1. K. Shinoda, T. Nakagawa, B. Tamamushi, and T. Isemura, *Colloidal Surfactants*, Academic Press, New York, 1963.
2. K. Shinoda and E. Hutchinson, J. Phys. Chem., *66*, 577 (1962).
3. K. Shinoda, J. Colloid Interface Sci., *34*, 278 (1970).
4. J. M. Corkill, J. F. Goodman, and S. P. Harrold, Trans. Faraday Soc., *65*, 287 (1969).
5. J. S. Clunie, J. F. Goodman, and P. C. Symons, Trans. Faraday Soc., *65*, 287 (1969).
6. M. J. Schick, J. Colloid Sci., *17*, 801 (1962).
7. E. H. Crook, G. F. Trebbi, and D. B. Fordyce, J. Phys. Chem., *68*, 3592 (1964).
8. J. E. Carless, R. A. Challis, and B. A. Mulley, J. Colloid Sci., *19*, 201 (1964).
9. J. M. Corkill, J. F. Goodman, and R. H. Ottewill, Trans. Faraday Soc., *57*, 1627 (1961).
10. K. W. Herrmann, J. Phys. Chem., *66*, 295 (1962).
11. H. S. Frank and M. W. Evans, J. Chem. Phys., *13*, 507 (1945).
12. K. Shinoda, J. Phys. Chem., *81*, 1300 (1977).
13. F. W. Fowkes, *Solvent Properties of Surfactant Solutions*, K. Shinoda, ed., Chap. 3, Marcel Dekker, New York, 1967.
14. R. C. Little and C. R. Singleterry, J. Phys. Chem., *68*, 3453 (1964).
15. R. C. Little, J. Colloid Interface Sci., *21*, 266 (1966).
16. L. Arkin and C. R. Singleterry, J. Colloid Sci., *4*, 537 (1949).
17. K. Shinoda and H. Arai, J. Colloid Sci., *20*, 93 (1965).
18. A. Kitahara, J. Phys. Chem., *69*, 2788 (1965).
19. H. Saito and K. Shinoda, J. Colloid Interface Sci., *35*, 359 (1971).

chapter ten

SOLUBILIZED SOLUTIONS

10.1 INTRODUCTION

In the field of solutions, one recurrent theme may be the attempt to improve the mutual solubility of oil and water, to overcome the difficulty characterized by the old saying, "Oil and water don't mix" [1,2].

Among the various ways of increasing this mutual solubility is the addition of a third substance. This addition may increase solubility by a number of different mechanisms, e.g., (1) co-solvency and blending, (2) solubilization, the formation of a pseudophase (micelle), a surfactant phase, or a liquid crystalline phase by the additive which has solvent properties different from the original solution, (3) melting the solute by the formation of a solute-additive mixture, and (4) formation of soluble complexes.

1. *Co-solvency and blending*. These are the terms associated with systems in which comparable amounts of two different solvents are employed to increase the solubility of solute. Unlike solubilization, the effect is not remarkable unless fairly large amounts of the second solvent (the third substance) are added.

It is well known that a mixture of two indifferently effective solvents may have solvent properties far exceeding those of the individual solvents. For example, mixtures of alcohol and tetrachloroethane readily dissolve cellulose acetate, although neither

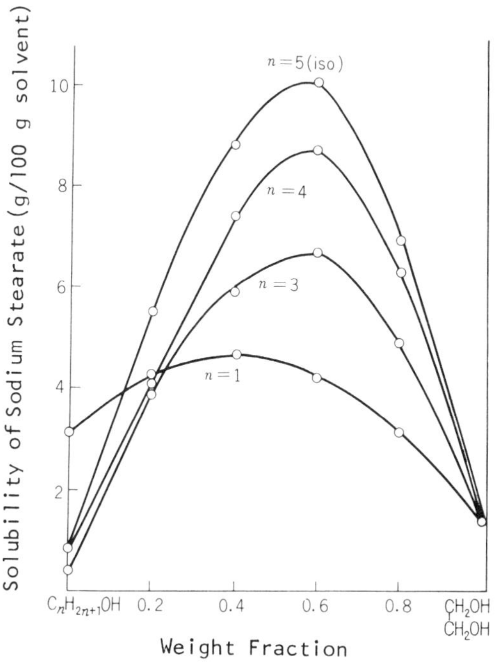

Fig. 10.1 Solubility of sodium stearate at 25°C in mixtures of ethylene glycol and methanol (n = 1); ethylene glycol and propanol (n = 3); ethylene glycol and butanol (n = 4); and ethylene glycol and isopentanol (n = 5).

liquid alone is a particularly good solvent for this polymer.

Palit [3] tried to find mixed solvents in which one component combines with the substance to be dissolved, while the other component dissolves the combination. Thus, a mixture of a glycol and a hydrocarbon may be used to dissolve soaps which are highly insoluble in either at room temperature. As much as 40 g of magnesium laurate may be dissolved in 100 g of a 1:4 mixture of propylene glycol and chloroform, although neither solvent alone will dissolve as much as 1.5 g/100 g of solvent. Palit suggests that the glycol associates with the ionic head of the soap molecule and that the hydrocarbons, chlorinated hydrocarbons, esters, and the like dissolve the paraffin chain of the soap. The solubility of sodium stearate at 25°C in mixtures of ethylene glycol and various alcohols is shown in Fig. 10.1.

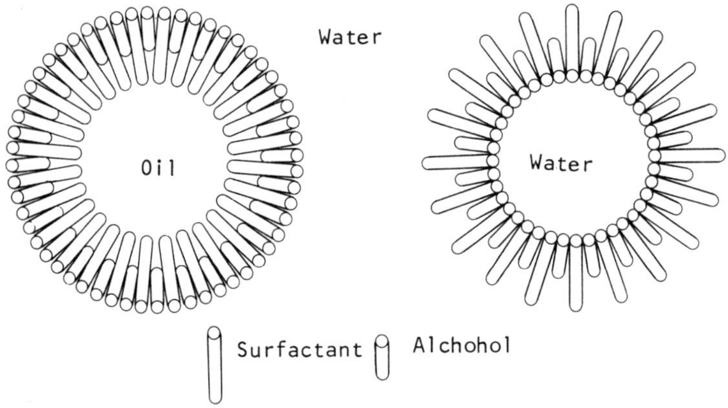

Fig. 10.2 Solubilization of alcohol and nonpolar oil (or water) by a spherical micelle.

M. E. L. McBain and Hutchinson have written [4], "We may regard these systems as examples of the use of soaps as means of bringing about the mutual solution of a glycol and a hydrocarbon." J. W. McBain used the term "blending" to describe this type of action.

2. *Solubilization.* Since the critical concentration for the formation of micelles (cmc) is, in general, so low, most surfactants are in the micellar state even in dilute aqueous solution. The cmc of a typical ionic surfactant (dodecylbenzene sulfonate), for example, is about 0.02% by weight. Thus, about 90% of the surfactant in a 0.2% by weight solution is in the micellar state. Since micelle formation is similar to a phase separation (see Sec. 9.2), a surfactant solution is similar to a two-phase mixture consisting of a water phase and a surfactant phase. The interior of the micelle is similar to liquid paraffin, so that added hydrocarbons dissolve preferentially in the micellar phase. For this reason, fairly dilute surfactant solutions will dissolve a certain amount of oil. In this respect, the phenomenon of solubilization differs from co-solvency. Polar oils, such as higher alcohols, dissolve in the palisade layer of the surfactant micelle, with the polar group oriented to the aqueous phase, as shown in Fig. 10.2 [4]. Phenol and ethylene glycol will preferentially dissolve in the hydrophilic polyoxyethylene

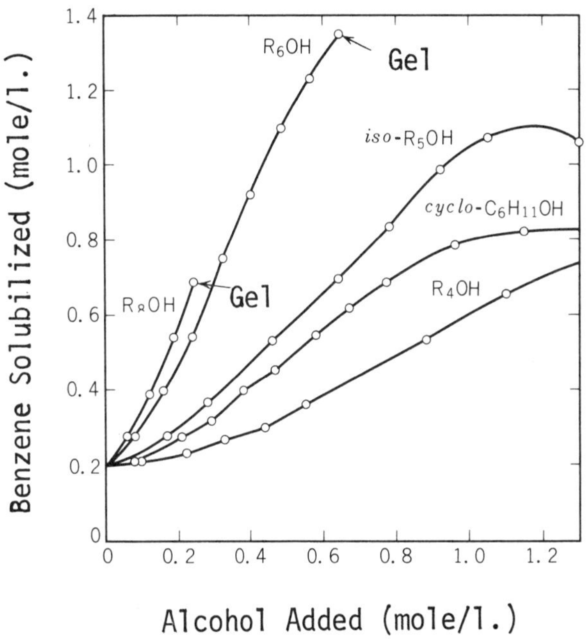

Fig. 10.3 Effect of added alcohols on the solubilization of benzene in a 0.285 mol/liter solution of potassium dodecanoate. From Ref. 7. Reprinted with permission.

core of a nonionic surfactant.

In other words, the principal location of solubilizates in or on the micelle depends on the nature of the solubilizate. The amount of oil solubilized is roughly proportional to the amount of micellar surfactant. As water is the most economical and widely obtainable solvent, the use of solubilization phenomena, for example, the extraction of oil-soluble materials into aqueous surfactant solutions, is widespread and important.

On the one hand, we can think of higher alcohols as being solubilized in the micelle, but on the other hand, it is possible to think of them as substituting for some surfactant molecules in the solubilization of nonpolar oils. In fact, the solubilization of hydrocarbons is enhanced by the addition of alcohols, amines, and mercaptans [5,6]. An example of this is shown in Fig. 10.3 [7].

In order to mutually dissolve oil and water with as small an amount of surfactant as possible, it is important to determine those factors which increase solubilization, and to establish a guiding principle which will serve to identify the surfactant appropriate to a given oil-water system.

Solubilization of oil and water through the formation of a liquid crystalline or surfactant phase is an even more efficient method of dissolving oil and water. In order to form a surfactant (or liquid crystal) phase, the hydrophilic and lipophilic properties of the surfactant or surfactant blend (mixture of a hydrophilic and a lipophilic surfactant) must be well balanced. In nonionic surfactant systems, the hydrophile-lipophile property will balance at an optimum temperature (PIT), since the balance shifts to lower hydrophilicity at higher temperatures, whereas in a system consisting of a ionic surfactant plus a cosurfactant, the system will be balanced at some optimum mixing ratio [8].

10.2 SOLUBILIZATION OF OIL IN AQUEOUS SOLUTION OF NONIONIC SURFACTANT

Change in temperature does not significantly change solubilization in aqueous solutions of ionic surfactants. However, the amount of oil solubilized by an aqueous solution of a nonionic surfactant changes substantially as a function of temperature. Consequently, the inclusion of measurements on temperature variation is indispensable. The solubilization of n-heptane, cyclohexane, and ethylbenzene in a 1.0% aqueous solution of polyoxyethylene (9.2) nonylphenol ether, $C_9H_{19}C_6H_4O(CH_2CH_2O)_{9.2}H$, is shown as a function of temperature in Fig. 10.4 [9].

It is evident from Fig. 10.4 that a turbidimetric measurement on a solution of nonionic surfactant does not suffice for the determination of the solubilization endpoint, since the turbidity may arise either because the limit of micellar solubilization has been reached, or because of the depression of the cloud point due to the addition of solubilizate. Beyond the solubilization limit, excess

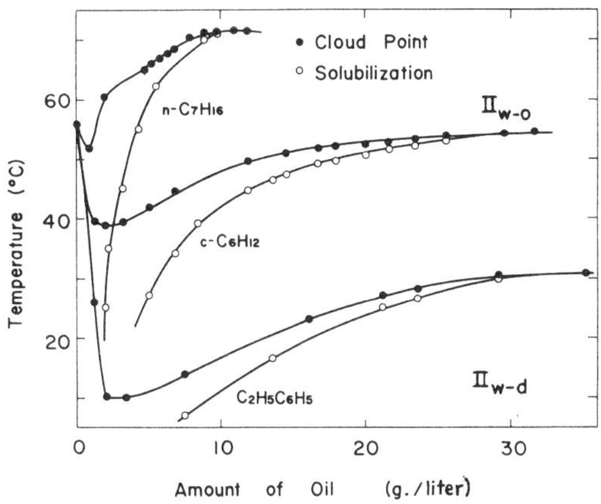

Fig. 10.4 The solubilization of n-heptane, cyclo-hexane, and ethyl benzene in a 1% aqueous solution of $R_9C_6H_4O(CH_2CH_2O)_{9.2}H$. Filled circles are cloud points, and surfactant phase separates from aqueous solution above this temperature. Open circles are the solubilization end-point; excess hydrocarbon separates from the micellar solution below this temperature or at greater solubilizate concentrations. From Ref. 9. Reprinted with permission.

oil disperses into the aqueous solution, forming an O/W-type emulsion. In the figure, the region II_{w-o} is a two phase solution, in which the oil phase is dispersed in the aqueous phase. The curve which links the open circles illustrates the temperature dependence of solubilization of the various oils. The amount of hydrocarbon solubilized increases markedly with temperature. Although the solution is one phase, the scattering of light increases progressively with the amount of solubilizate, as the swollen micelles become bigger. Thus, the solution changes continuously into the so-called microemulsion state. In this state, this solution is turbid but infinitely stable. If the temperature of the solubilized solution is raised to the cloud point, the solution splits into aqueous and surfactant phases. The cloud point curve is shown by the closed circles. The region above the cloud point, where the system has become two-phase, is designated II_{w-d}.

Why does solubilization markedly increase close to the cloud point? It is known from light-scattering studies that the micellar

Table 10.1 Cloud Points of 1 wt% Aqueous Solutions of Polyoxyethylene (9.2) Nonylphenylether Saturated with Various Oils [10]

Type of oils	Cloud point (°C)	Type of oils	Cloud point (°C)
Liquid paraffin	80.4	No oil	56
n-$C_{16}H_{34}$	80	c-C_6H_{12}	54
n-$C_{12}H_{26}$	79.3	CCl_2CCl_2	31
n-$C_{10}H_{22}$	79	$C_2H_5C_6H_5$	30.5
n-C_7H_{16}	71.5	C_6H_6	Below 0

aggregation number of a surfactant increases rapidly near the cloud point [10]. If this is the case, the solubilization of oil will also increase. Hence, the oil-saturated cloud point is an important characteristic temperature for a given oil-water-surfactant system. This temperature varies markedly with the types of oil (Table 10.1) and with the oxyethylene chain length (Table 10.2).

When the solution splits into two phases above the cloud point, solubilization does not occur. However, no separation of oil phase takes place, since the oil dissolves in the surfactant phase. In other words, the solubility of oil in a two-phase system (water plus surfactant phase) does not decrease above the cloud point. Accordingly, detergent action, or the ability to extract oil from a system, does not decline. If the temperature is further raised, however, the solubility of water in the surfactant phase decreases,

Table 10.2 Relation Between Hydrophilic Chain-lengths and the Cloud Points of 5 wt% Surfactant in Water-Cyclohexane

Surfactant (pure)	Cloud point (°C)	Surfactant (commercial)	Cloud point (°C)
$C_{12}H_{25}O(CH_2CH_2O)_6H$	27	$C_9H_{19}C_6H_4O(CH_2CH_2O)_{6.2}H$	~10
$C_{12}H_{25}O(CH_2CH_2O)_7H$	39	$C_9H_{19}C_6H_4O(CH_2CH_2O)_{7.4}H$	44
$C_{12}H_{25}O(CH_2CH_2O)_8H$	50.5	$C_9H_{19}C_6H_4O(CH_2CH_2O)_{9.7}H$	61
		$C_9H_{19}C_6H_4O(CH_2CH_2O)_{17.7}H$	~100

and the surfactant phase changes to an oil phase. That is, the surfactant, which has formed the continuous phase at the cloud temperatures, dissolves in the oil, forming oleophilic micelles, at the higher temperature.

10.3 SOLUBILIZATION OF WATER IN HYDROCHARBON SOLUTIONS OF NONIONIC SURFACTANT

The amount of water solubilized in a hydrocarbon (including halogenated hydrocarbons) solution of nonionic surface-active agent also changes strikingly as a function of temperature, so that measurements of temperature variation are necessary [11]. For example, the solubilization of water in cyclohexane containing 10 wt% of polyoxyethylene (9.6) nonylphenyl ether is shown in Fig. 10.5.

The open circles indicate the solubilization limit of water, while the closed circles show the cloud point. In contrast to an aqueous system, the surfactant phase separates from the oil phase

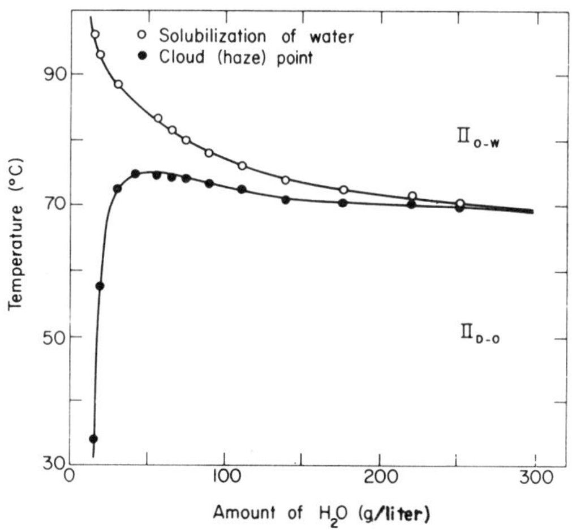

Fig. 10.5 Solubilization of water in cyclohexane containing 10 wt% polyoxyethylene (9.6) nonylphenyl ether. Open and filled circles represent solubilization end-point and cloud point, respectively.

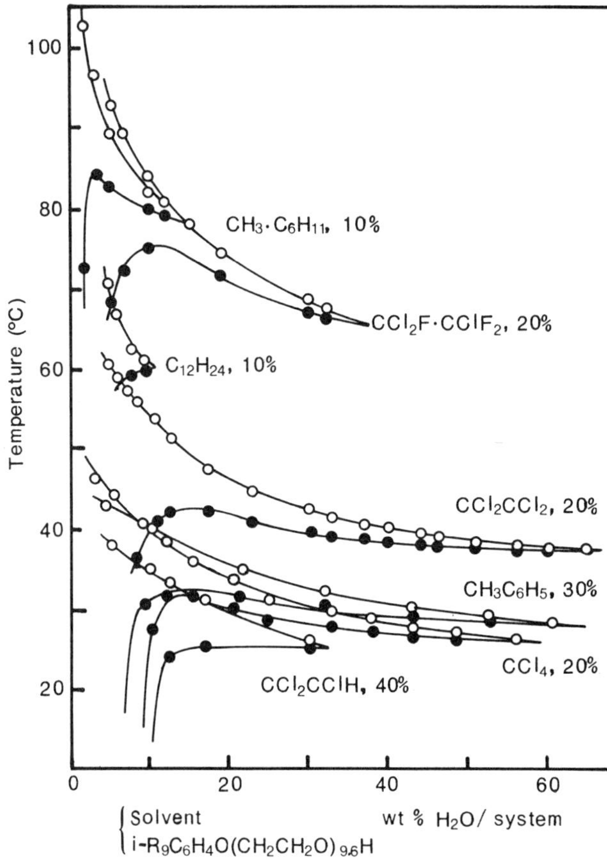

Fig. 10.6 Solubilization of water in various solvents containing polyoxyethylene (9.6) nonylphenyl ether. Concentrations of surfactant (wt%) are indicated in the figure. From Ref. 11. Reprinted with permission.

below the cloud point temperature. The solubilization region between the cloud point curve and solubilization curve is a transparent phase in which the water-swollen oleophilic micelles are dispersed in oil. The amount of light scattered increases continuously with the amount of water solubilized, and becomes quite turbid close to the cloud point. However, the solution is still one thermodynamically stable phase. Beyond the solubilization limit, however, the excess water splits into two phases, the type of dispersions being

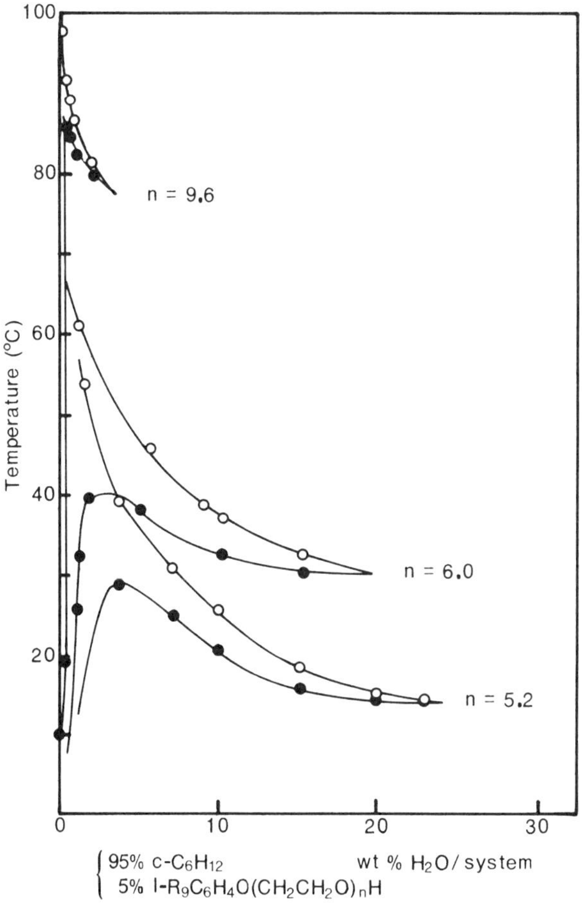

Fig. 10.7 Effect of the hydrophilic chain length of the surfactant on solubilization of water in cyclo-hexane containing 5 wt% of surfactant. n = polyoxyethylene chain length. From Ref. 11. Reprinted with permission.

w/o. The region II_{o-w} in the figure corresponds to a two-phase region, with water splitting out from the oil phase. The optimum temperature for solubilization of water in nonaqueous solutions of nonionic surfactants also changes with the type of oil and kind of surfactant, as illustrated in Figs. 10.6 and 10.7 [11].

Since solubilization changes so markedly with temperature, the type of oil or of surfactant, it is important that the surfactant

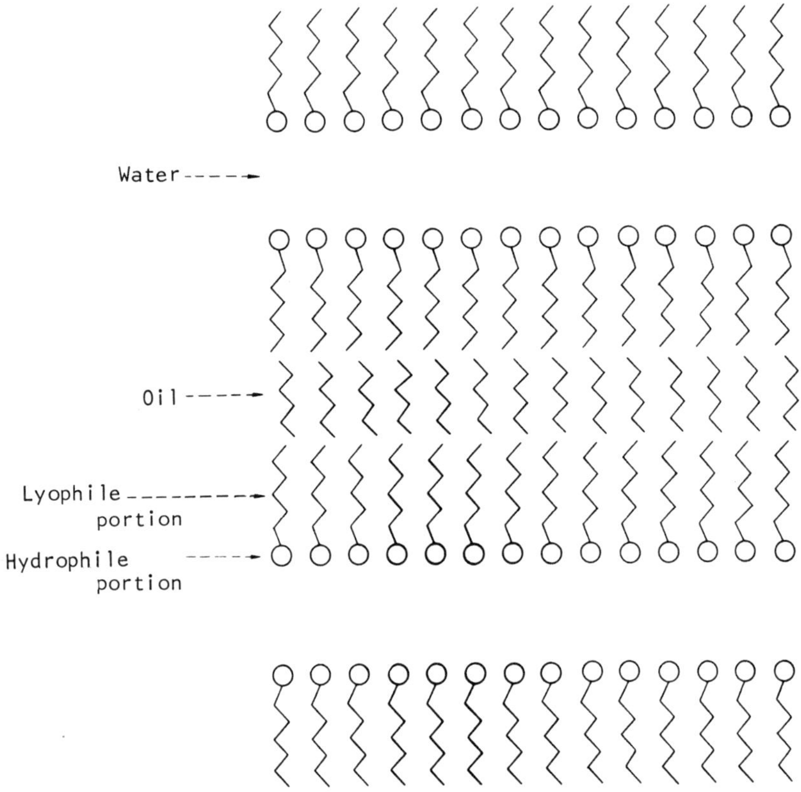

Fig. 10.8 Schematic representation of surfactant phase, showing oil and water dissolved between the surfactant layers.

chosen be appropriate for the given oil at the given temperature.

Let us now consider why the solubilization increases markedly at a certain temperature. Imagine the hydrophilic surface of an aggregated surfactant monolayer in contact with water and the lipophilic surface in contact with hydrocarbon. If the interfacial tension between water and the hydrophilic surfactant surface is smaller than that between hydrocarbon and the lipophilic surface of the surfactant monolayer, the hydrophilic surface will expand and the lipophilic surface will contract. As a result, the surfactant monolayer will be convex with respect to the water phase. Under these circumstances, the surfactant will disperse in water as micelles containing solubilized oil. If the temperature of the solution is raised, however, the interaction between water and the hydrophilic portion of the surfactant decreases, and the surfactant disperses in the oil, forming oleophilic water-swollen micelles. If the temperature is at some median value, the concave and convex tendencies of the surfactant monolayer just balance, and the surfactant aggregates infinitely, e.g., a surfactant phase separates at the median temperature. This surfactant phase may have the layered structure shown schematically in Fig. 10.8. Water and oil are highly soluble in the surfactant layers. Emulsions and micelles are found in the inverse forms above and below the temperature at which the surfactant phase separates, i.e., phase inversion of the emulsion occurs. Hence, the aggregation number of micelles and the solubilization of water increases close to the phase inversion temperature (PIT), i.e., at the cloud point.

10.4 PHASE DIAGRAM OF TERNARY SYSTEM WATER/HYDROCARBON/NONIONIC SURFACTANT

It is important to study the phase equilibria of ternary systems composed of water, hydrocarbon, and nonionic surfactant in order to understand the mutual relations involved in the solubilization of oil (or water) in aqueous (or nonaqueous) solutions, as well as emulsification, detergency, dispersion, and so on, by nonionic surfactants. There are many variables in such a study, such as the types of oils used, the kinds of surfactants, the component composition, and the temperature. Among saturated hydrocarbons, the behavior of cyclohexane may be regarded as typical. Although the water-oil composition has to be explored over the entire volume fraction range, as a practical matter the concentration of the surfactant may be fixed. The effect of temperature is important, since, as we have seen, the amount of solubilization, the type and stability of emulsions, the dispersion state of surfactants, etc., depend so markedly on temperature, while the effect of a rise in temperature corresponds effectively to a decrease in the hydrophilic

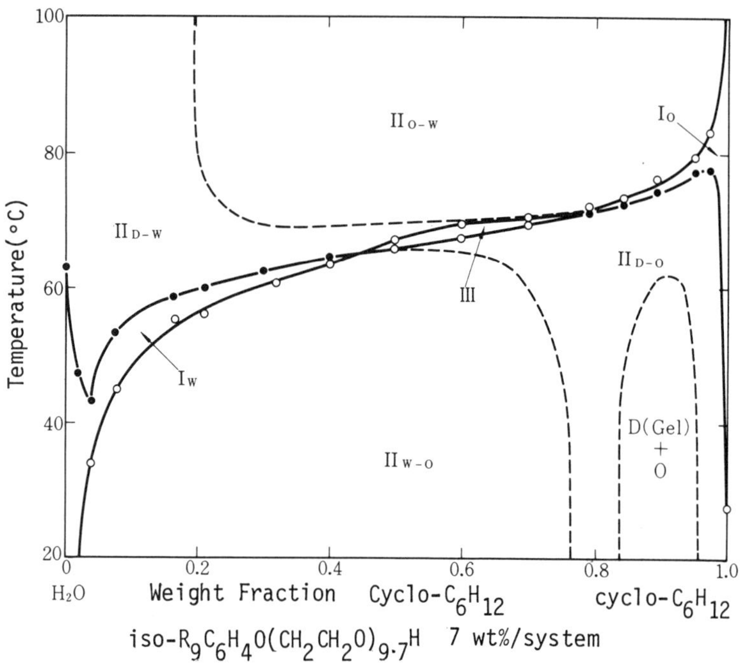

Fig. 10.9 Phase diagram of the system water-cyclohexane containing 7 wt% of polyoxyethylene (9.7) nonylphenyl ether. From Ref. 12. Reprinted with permission.

chain length of a nonionic surfactant, and thus affords information on the effect of change in oxyethylene chain length. A change in either the hydrophilic or lipophilic chain length in the surfactant shifts the phase equilibria either to higher or lower temperatures, but the pattern is similar in both cases. Thus, the phase equilibria and the dispersion types in a water-cyclohexane system containing 7 wt% on the system of polyoxyethylene (9.7) nonylphenyl ether are plotted in Figs. 10.9 and 10.10 as a function of temperature [12]. The curve connecting the open circles in Fig. 10.10 corresponds to the phase inversion temperature of the emulsion.

It is evident from Fig. 10.10 that the PIT is very close to the temperature at which the solubilization of oil (or water) in an aqueous (or nonaqueous) solution of a nonionic surfactant is at a

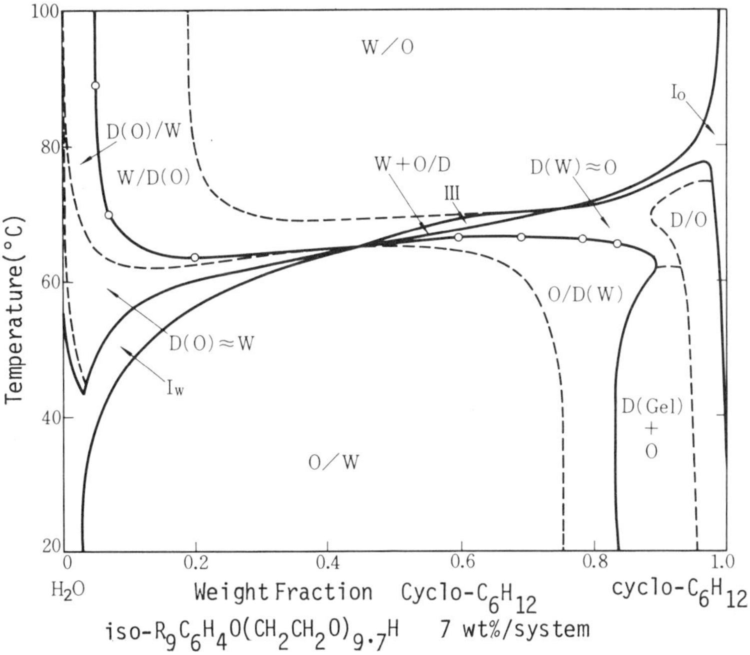

Fig. 10.10 Effect of temperature and composition on the dispersion types in a system composed of water, cyclohexane, and polyoxyethylene (9.7) nonylphenyl ether (7 wt%). From Ref. 12. Reprinted with permission.

maximum. The phase inversion temperature is the temperature at which the hydrophile-lipophile property just balances and is therefore an important characteristic property in surfactant selection.

The phase inversion temperature of nonionic surfactants changes with the kinds of surfactant and the types of oils. For example, the PITs of 1:1 mixtures of various hydrocarbons and water containing 1.5 wt% on the system of various kinds of nonionic surfactants are plotted in Fig. 10.11. The location of the surfactants on the abscissa is determined by situating the PITs for the n-heptane-water system on a straight line of slope unity [13].

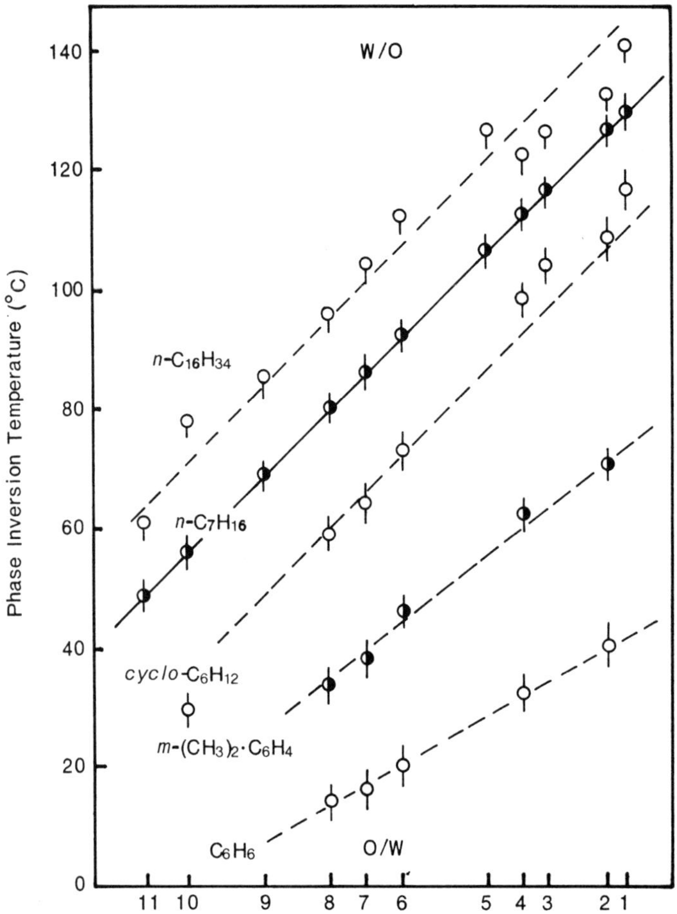

Fig. 10.11 Correlation among phase inversion temperatures for emulsions of different hydrocarbons. 1, Tween 40; 2, i-$R_9C_6H_4O(CH_2CH_2O)_{17.7}H$; 3, Tween 60; 4, i-$R_9C_6H_4O(CH_2CH_2O)_{14.0}H$; 5, $R_{12}O(CH_2CH_2O)_{10}H$; 6, i-$R_9C_6H_4O(CH_2CH_2O)_{9.6}H$; 7, i-$R_{12}C_6H_4O(CH_2CH_2O)_{9.0}H$; 8, i-$R_9C_6H_4O(CH_2CH_2O)_{7.4}H$; 9, Pluronic L-64; 10, i-$R_9C_6H_4O(CH_2CH_2O)_{6.2}H$; 11, Pluronic L-62. From Ref. 13. Reprinted with permission.

10.5 SOLUBILIZATION BY IONIC SURFACTANTS

A large amount of data on solubilization by ionic surfactants is available [1]. Solubilization of hydrocarbons in an ionic surfactant solution is very little dependent on the temperature, since the interaction between the ionic group of the surfactant and water is itself little affected by change in temperature. Thus, there is little significance in studies of the temperature effect of solubilization above the Krafft point.

As we have seen, however, increase in temperature or decrease in hydrophilic chain length have equivalent effects on solubilization

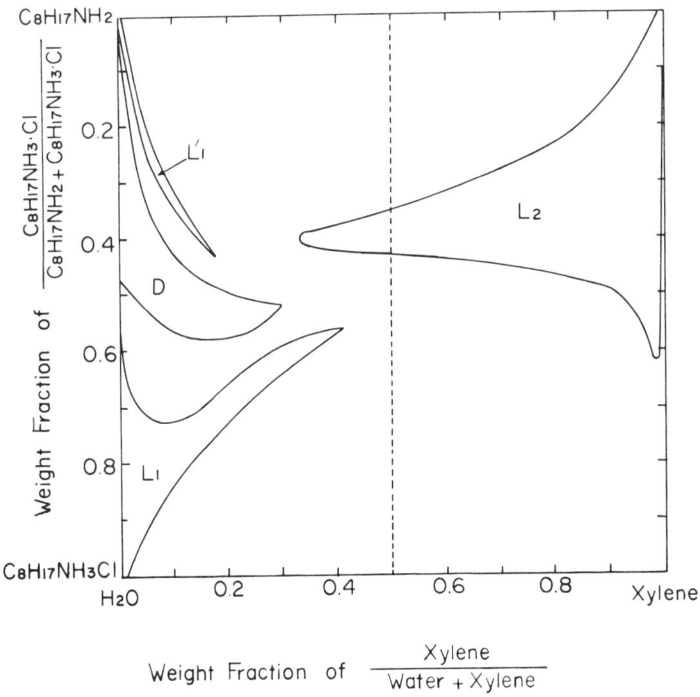

Fig. 10.12 Phase equilibria of octylamine, octyl ammonium chloride, water, and p-xylene at 25°C. Octylamine + octyl ammonium chloride and p-xylene + water were maintained at total concentrations of 20 wt% and 80 wt%, respectively. From Ref. 15. Reprinted with permission.

by nonionic surface-active agents. This implies that the solubilization behavior of an ionic surfactant will also change appreciably as a result of the partial substitution of a lipophilic surfactant. In fact, the solubilization increases markedly when the HLB of the surfactant mixture is optimum, i.e., the mixing ratio is optimum [8,14,15].

One example is shown in Fig. 10.12 [15]. This system contains a total of 20 wt% of octylamine and octylammonium chloride. Three liquid phases L_1, L', L_2 and one crystalline phase D, exist. Region L_2, nonaqueous phase, consists of amine dissolved in p-xylene. The maximum solubilization of water at 53 wt% of the system occurs at an octylamine/octylammonium chloride ratio of 1.5 [14]. In the region L_1, octylammonium chloride is dissolved in water and solubilized xylene. The maximum solubilization of xylene (33 wt%) extends up to an octylamine/octylammonium chloride ratio of 0.8 (w/w).

10.6 SOLUBILIZATION OF WATER AND OIL IN SURFACTANT: FORMATION OF SURFACTANT AND LIQUID CRYSTALLINE PHASES

In addition to the solubilization of oil (water) in aqueous (nonaqueous) surfactant solutions, there exist other types of dissolution (or solubilization) of both oil and water in surfactant phases [12,16] and in the formation of liquid crystalline phases [16-18]. The distinct feature of this type of solubilization resides in the large amount of solubilization per mole of surfactant which may be achieved, although at the expense of a high concentration of surfactant, i.e., 5 to 10 wt%.

An example of solubilization in the surfactant phase is shown in Fig. 10.13. The surfactant phase (the island close to the water-hexadecane axis) was observed in a composition containing 10 wt% of surfactant, 45 wt% of water, and 45 wt% of hexadecane at a temperature of 25°C.

Formation of liquid-crystalline phases [17] in the system consisting of water, sodium decanoate, and decanol is shown in Fig. 10.14. The liquid crystalline phases B and D, shown in Fig. 10.14,

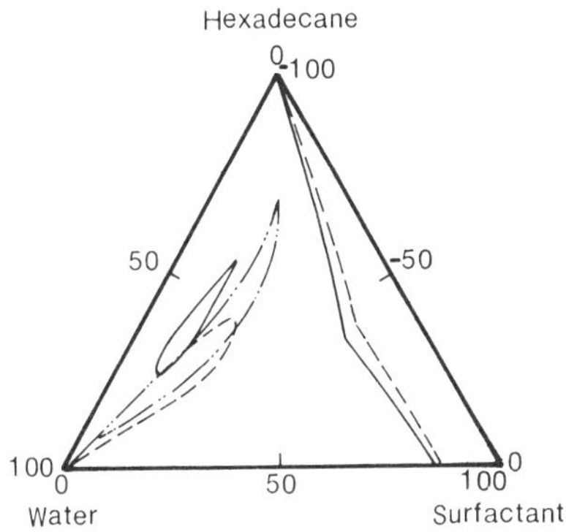

Fig. 10.13 Three-component phase diagram for water, tetraoxyethylene dodecyl ether, and hexadecane at 25°C (solid line), at 20°C (dotted line), and at 18.5°C (broken line). From Ref. 16. Reprinted with permission.

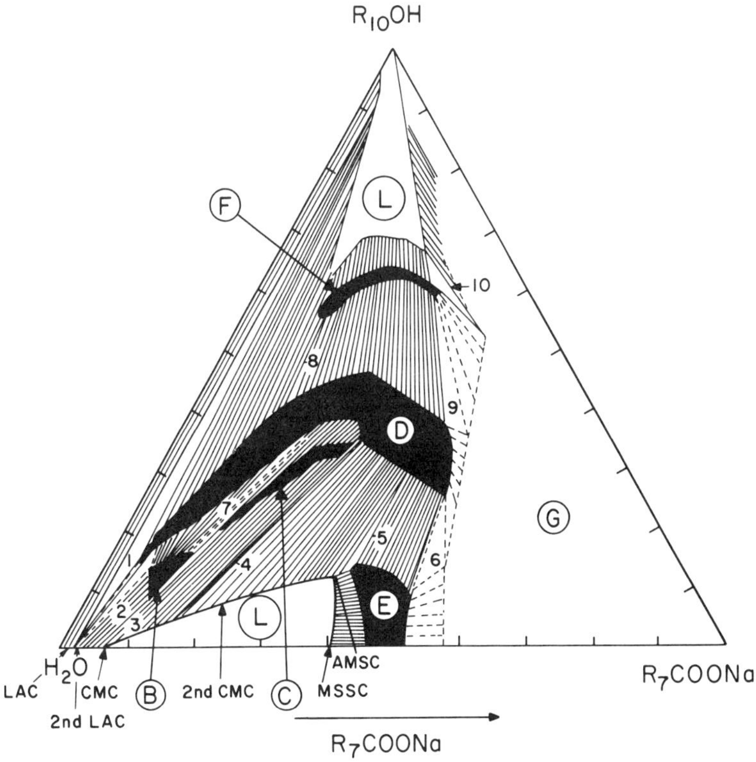

Fig. 10.14 Phase diagram for the three-component system sodium caprylate-decanol-water at 20°C. Concentrations are given in weight percent.

contain large amounts of water and decanol. When the decanol is replaced by a hydrocarbon, there exists no comparable liquid-crystalline phase containing a large amount of water or oil. Thus, an ionic surfactant which is usually strongly hydrophilic requires a co-surfactant (such as decanol) in order to solubilize a large amount of oil or water.

REFERENCES

1. M. E. L. McBain and E. Hutchinson, *Solubilization and Related Phenomena*, Academic Press, New York, 1955.
2. K. Shinoda (ed.), *Solvent Properties of Surfactant Solutions*, Marcel Dekker, New York, 1967.
3. S. R. Palit, J. Indian Chem. Soc., *19*, 271 (1942); J. Am. Chem. Soc., *69*, 3120 (1947).
4. W. D. Harkins, R. W. Mattson, and R. Mittelman, J. Chem. Phys., *15*, 763 (1947).
5. H. B. Klevens, J. Chem. Phys., *17*, 1004 (1949).
6. H. B. Klevens, J. Am. Chem. Soc., *72*, 3581 (1950).
7. K. Shinoda and H. Akamatsu, Bull. Chem. Soc. Japan, *31*, 497 (1958).
8. G. Gillberg, H. Lehtinen, and S. Friberg, J. Colloid Interface Sci., *33*, 40 (1970).
9. H. Saito and K. Shinoda, J. Colloid Interface Sci., *24*, 10 (1967).
10. T. Nakagawa, K. Kuriyama, and H. Inoue, J. Colloid Sci., *15*, 268 (1960).
11. K. Shinoda and T. Ogawa, J. Colloid Interface Sci., *24*, 56 (1967).
12. K. Shinoda and H. Saito, J. Colloid Interface Sci., *26*, 70 (1968).
13. K. Shinoda and H. Arai, J. Phys. Chem., *68*, 3485 (1964).
14. S. Friberg, Kolloid-Z.Z. Polym., *244*, 333 (1971).
15. S. I. Ahmad, K. Shinoda, and S. Friberg, J. Colloid Interface Sci., *47*, 32 (1974).

16. S. Friberg and I. Lapczynska, Prog. Colloid Polymer Sci., *56*, 16 (1975).

17. P. Ekwall, I. Danielsson, and L. Mandell, Kolloid-Z., *169*, 113 (1960).

18. P. Ekwall, L. Mandell, and K. Fantell, *Molecular Crystals and Liquid Crystals*, Vol. 8, Gordon and Breach, New York, 1969, pp. 157-213.

chapter eleven

FACTORS WHICH CONTROL SOLUBILITY (SUMMARY)

11.1 INTRODUCTION

The ideal solution discussed in Chap. 2 is an idealized solution, i.e., no change in volume, zero heat of solution, and the size of the molecules of each component is equal. If we view solutions in general, we recognize that this is a very special solution. On the other hand, a regular solution is more general in that one can explain its behavior in a general way, irrespective of whether the solute is solid, gas, or liquid, and whether the molecular sizes are different, and the heat of solution is large. It fails in special situations, i.e., when dipole effects are important, when association or reaction takes place, etc.

Complete random mixing is the only assumption in regular solution, provided the energy of mixing is evaluated experimentally either from the partial pressure or from the solubility of the solute. As this assumption is quite reasonable in nonassociating or nonsolvating systems, the concept of regular solutions can be applied to the study of solubility in many systems.

However, there are many practical cases where a simple application of the theory of regular solutions may not suffice, since the systems in question may contain water which has a strong tendency to associate, or a molecule which has a strong tendency to orient or to other types of specific interaction. The most striking feature of these solutions may be the dissolution due to the orientation, arrangement, conformation, and structure formation of solute molecules otherwise practically insoluble by random mixing only. Although the saturation concentration of singly dispersed species can be interpreted by the concept of the regular solution, we may need to consider an entirely new mechanism to understand dissolution phenomena as a whole.

We shall discuss, step by step, the methods one may use to increase or decrease solubility in a given situation. This chapter constitutes a summary of Chaps. 1 to 10 and is organized to be used in the solution of practical solubility problems.

11.2 FACTORS THAT CONTROL SOLUBILITY IN REGULAR SOLUTIONS

The entropy of solution, the enthalpy of solution, and the Gibbs energy associated with the solution are the three important factors controlling the solubility. The solubility is expressed by the following equation:

$$\ln a_2 = \ln x_2 + \frac{V_2 \phi_1^2 (\delta_1 - \delta_2)^2}{RT} \tag{4.10}$$

Here, a_2 is the relative activity of the solute relative to the pure liquid state, x_2 is the solubility expressed as a mole fraction, V_2 is the molar volume of the solute, ϕ_1 is the volume fraction of solvent, and δ_1 and δ_2 are the solubility parameters of solvent and solute, respectively. If $(\delta_1 - \delta_2)^2$ or $V_2(\delta_1 - \delta_2)^2$ is determined from the vapor pressure, or by the direct measurement of the solubility instead of using solubility parameters, Eq. (4.10) is more accurate and can be applied to more general systems. In that case, the solubility is expressed as:

$$\ln a_2 = \ln x_2 + \frac{B \phi_1^2}{RT} \tag{4.10'}$$

With this in mind, we can describe methods suitable for increasing solubility of certain substances:

1. *Reduce the heat of solution by appropriate choice of solvent.* To achieve this, select a solvent whose solubility parameter is close to that of the solute. As the solubility parameters of similar molecules are generally of the same magnitude, their mutual solubility is good. This corresponds to reducing the value of $(\delta_1 - \delta_2)^2$ in Eq. (4.10), resulting in an increase in solubility, x_2.

It should be remembered that the solubility parameters are not accurate in estimating the heat of solution unless the only forces on the molecules under consideration are solely due to London dispersion forces. The presence of a large number of methyl groups may induce abnormal behavior, even when such groups as —OH, —SH, $—NH_2$, and —COOH are absent. For example, the relationship between the solubility of iodine in various regular solvents and $(\delta_1 - \delta_2)^2$ is in good agreement with theory (solid line in Fig. 11.1), but large deviations from theory are evident with those solvents containing several $—CH_3$ groups [1]. The filled circles represent aromatic hydrocarbons, where interactions of the electron transfer type with iodine contribute to the significant deviations.

The cohesion owing to dipole interaction is strong between molecules with strong dipoles. On the other hand, there is no such interaction between molecules with dipole and no dipole; hence, the interaction between molecules of different species becomes less than the geometrical average and the heat of solution becomes greater, thus reducing their mutual solubility. The empirical rule that "a polar substance is more soluble in a polar solvent, and a nonpolar substance more soluble in a nonpolar solvent" is thus rationalized.

2. *Select a solute of small molar volume.* If the solvent is given, and a choice of solutes of similar nature is possible, selection of a solute of small molar volume will result in increased solubility. This point is made clearer by writing Eq. (4.10') in the form

$$\ln a_2 = \ln x_2 + \frac{B' V_2 \phi_1^2}{RT} \tag{11.1}$$

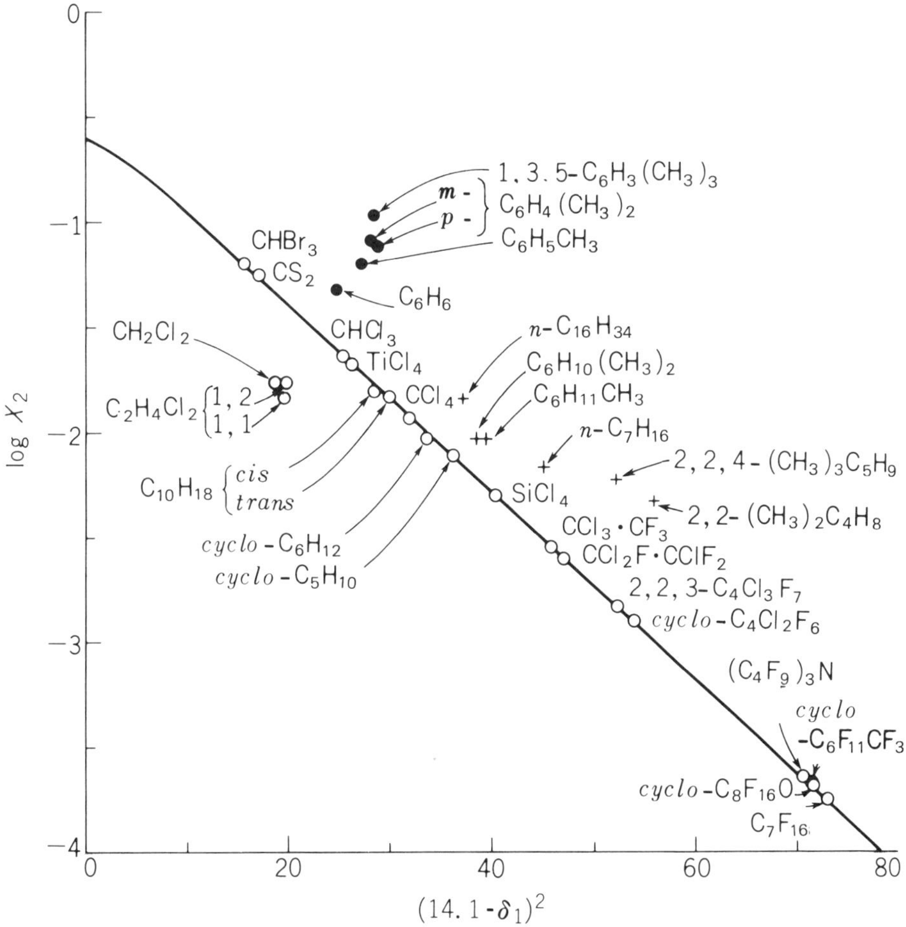

Fig. 11.1 Relation between the solubility of iodine and the solubility parameters of the solvents. Deviations occur for solvents containing π-electrons, methyl groups, and dipoles. From Ref. 1(b). Reprinted with permission. Copyright by the American Chemical Society.

where B' is the heat of solution per unit volume of solute. Since the relation between heat of solution and solubility is logarithmic, the effect is significant.

3. *Select a solvent of small molar volume.* It may appear that the solubility is independent of the molar volume of the solvent,

since this term does not appear explicitly in Eq. (4.10). However, the practical unit of solubility is the amount dissolved in a given volume, and not the mole fraction used in solution theory. If the solubilities, expressed in mole fraction, are equal, the amount of solute dissolved in a given volume will be greater in the solvent of smaller molar volume. The effect is proportional to the ratio of molar volumes.

4. *Increase the relative activity of the solute.* (a) To increase the solubility of a gas, increase its pressure. The relative activity a_2 increases approximately in proportion to the pressure, and the solubility x_2 also increases proportionally. (b) Select a solute of low melting point, or lower the melting point by other choice. The solubility of a solid decreases more rapidly with decreasing temperature than does a supercooled liquid, because the relative activity of a solid a_2^s decreases with temperature depression. In practice, one would seek to replace the original solute with one of similar properties, but lower melting point. For example, the solubilities of the similar materials, phenanthrene and anthracene, are compared in Table 11.1 [2], and the three isomers of dinitrobenzene which exhibit markedly different solubilities in benzene and water, are shown in Table 11.2 [3].

Table 11.1 Influence of Melting Point upon Solubility, mol% at 25°C [2]

	Solubility	
	Phenanthrene	Anthracene
Boiling point (°C)	340	342
Melting point (°C)	99.6	218
Hexane	4.2	0.18
Ethyl ether	15.1	0.59
Carbon tetrachloride	18.6	0.63
Benzene	20.7	0.81
Carbon disulfide	25.5	1.12
Ethanol	1.25	0.09
Ideal	22.1	1.07

Table 11.2 Influence of Melting Point upon Solubility

Solute	Melting point (°C)	Solubility: in C_6H_6 mol% at 50°C [2]	Solubility: in H_2O g/liter at 25°C [3]
m-$C_6H_4(NO_2)_2$	90	37.6	0.5328
o-$C_6H_4(NO_2)_2$	116	17.5	0.1247
p-$C_6H_4(NO_2)_2$	170	3.1	0.0618

The difference between molecular shapes of phenanthrene and anthracene has no effect on the liquid properties, and hence the boiling points are very close. However, the melting points are very different, and the relative activity of phenanthrene is approximately 20 times greater than that of anthracene at 25°C. Since there is little difference in the intermolecular forces, it is evident that the difference in solubility is entirely due to the difference in melting points. Similarly, PTFE is difficult to dissolve because of its high melting point and very large heat of fusion.

5. *Increase the solubility of a solid or liquid by raising the temperature.* As the temperature increases the solubility parameter decreases, and also the solubility parameter difference. Consequently, the heat of solution decreases, and, since the enthalpy term in Eq. (4.10) is inversely proportional to the temperature, the solubility increases. This applies also to solubility of liquids. In the case of a solid, there is the additional effect of the increase in the relative activity of the solid with increasing temperature. In some special cases, the heat of solution is negative owing to the specific interaction or to iceberg formation and the magnitude increases at lower temperature. In these instances, solubility increases with *decreasing* temperature. For example, in nonionic surfactants or organic amines versus water system, critical miscibility occurs at a lower consulate temperature (see Figs. 5.10 and 9.4).

6. *Increase the solubility of a gas by lowering the temperature.* In the case of a gas dissolved in a liquid, the logarithm of the relative activity of the gas may increase faster on lowering the temperature than the heat of solution increases. Hence, the solubility generally increases in some extent as the temperature decreases. However, the change in solubility is usually small. The extent and sign of the change can be predicted from Fig. 5.4, if the solubility parameter of the solvent in which the gas is dissolved is known.

11.3 GENERAL VIEWPOINTS ON SOLUBILITY

1. *Effect of hydrophilic groups.* In order to make an organic compound more water soluble, pH may be changed. The types of gegenions sometimes have a marked effect, because the melting point (and therefore relative activity of solute) changes with the types of gegenions. If chemical treatment of the solute is permissible, dissociable or hydrophilic groups can be attached to the molecule to make it more water soluble. The effect of hydrophilic groups on the solubility of an organic compound is illustrated in Table 11.3. As can be seen, the solubility increases markedly as we progress from nitrobenzene to aniline to phenol. The size and shapes of these molecules are similar and their boiling points do not differ very much. The difference in solubility must therefore be attributable to hydrogen bonding, $-NO_2$ forming weak hydrogen bonds, while $-NH_2$ forms a fairly strong bond, and $-OH$ the strongest.

Table 11.3 Solubility in Water of Benzene and its Derivatives at 20°C

Solutes	Solubility (mol%)	Dipole moment (Debye units)	Boiling point (°C)	Melting point (°C)
C_6H_6	0.019 (22°C)	0	80.1	5.5
$C_6H_5NO_2$	0.028 (20°C)	4.2	210.9	5.7
$C_6H_5NH_2$	0.65 (20°C)	1.5	184	-6.2
C_6H_5OH	1.27 (16°C)	1.7	181.4	41

Table 11.4 Surface Tension and Solubility Parameters of Halogenated Compounds

Substances	Surface tension, 20°C (dyn/cm^2)	Solubility parameter
C_7F_{16}	13.6	6.0
n-C_6H_{14}	18.4	7.3
$SiCl_4$	-	7.6
c-C_6H_{12}	25.3	8.2
CCl_4	27.0	8.6
C_6H_6	29	9.2
$CHCl_3$	27	9.2
CH_2Cl_2	26.5	9.8
$CHBr_3$	41.5	10.5
Br_2	41.5	11.5
CH_2I_2	50.8	11.8
I_2	(solid)	14.1

2. *Comparison of fluorides, chlorides, bromides, and iodides.* In comparing the halogens, it should be noted that the number of outer electrons increases in the order F, Cl, Br, I, and that these electrons are more loosely bound, since they are further from the nucleus. For this reason, the dispersion forces are weak for fluorine and fluoride, but as we move to bromides and iodides, the polarizability becomes greater and intermolecular forces stronger. As shown in Table 11.4, fluorocarbons exhibit the smallest surface tension among liquids at room temperature. On the other hand, liquids such as CH_2I_2 show higher surface tensions.

If the hydrogen of a hydrocarbon is replaced by iodine, the intermolecular forces become greater and the boiling point higher. In the case of substitution by fluorine, however, the intermolecular force decreases, but the boiling point remains almost unaffected, owing to the increase in molar volume.

3. *Solubility of isomer/homologue mixtures.* In the case when a mixture of homologues or isomers of hydrocarbons (or alcohols) is

shaken with water, the solubility changes with the amount of excess solute phase, because the composition of the excess solute in equilibrium with the solution changes.

For example, assume that the solute (component 2) is a mixture of two substances A and B whose mole fractions are, respectively, x_A and x_B, and whose respective solubilities in the solvent are S_A and S_B. When a sufficiently large amount of the mixture is shaken with water, the quantity dissolved is given by

$$S_A x_A + S_B x_B = S_2 \tag{11.2}$$

because isomers or homologous mixtures separated from water may form an ideal solution. The mole ratio of A and B in the solution is given by $S_A x_A / S_B x_B$. On the other hand, if the solubility is determined when a minute amount of excess solute has appeared, the mole ratio of A and B in the solution is x_A/x_B. Thus, the composition in the solute phase at the moment when separation begins is $x_A S_B / x_B S_A$. Thus, the mole fractions of A and B in the separated phase are:

$$x_A' = \frac{x_A S_B}{x_A S_B + x_B S_A}; \qquad x_B' = \frac{x_B S_A}{x_A S_B + x_B S_A} \tag{11.3}$$

Inserting x_A' and x_B' in Eq. (11.2), the amount dissolved is seen to be

$$S_2' = \frac{S_A S_B}{x_A S_B + x_B S_A} \tag{11.4}$$

Since the less soluble component is always concentrated in the separated phase, the solubility is smaller in the latter case.

11.4 SOLUTION BY A PARTICULAR MECHANISM

1. *Solution of polymeric substances.* A solid (crystalline) polymer in equilibrium with a solvent does not dissolve. This is due to the fact that the quantity $B'V_2$ in Eq. (11.1) becomes very large and a_2^s very small; accordingly x_2 also becomes very small.

The heat of solution of polymer is very large, even though the heat of solution of per unit volume may be very small. On the other hand, when a polymer in contact with a solvent is in the liquid state, the solvent molecules dissolve well in the polymer, since the molar volume and the heat of solution of solvent are both small. In the case where solvent and polymer are not completely soluble in each other, the volume fraction of polymer in the solvent phase is very small, but the volume fraction of solvent in the polymer phase is usually far greater than that of polymer. Polymer and the so-called solvent are completely miscible when the solubility of solvent in polymer is infinite.

In other words, the process of solution of a polymer involves the solvent dissolving in the polymer, and in order for this to take place readily it is necessary that the polymer be in the liquid form when in contact with the solvent (the temperature required varying with the kinds of solvents employed) (see Chap. 8).

2. *Solution of colloidal surfactants.* Surfactant molecules aggregate into micelles above the saturation concentration of singly dispersed species. Since micelles are sufficiently small with respect to the wavelength of light, the solution does not appear turbid. Thus, the surfactant solution is transparent and the solubility quite large. This phenomenon of a remarkable increase in solubility owing to the dispersion of small particles, pseudo-phase dispersion, represents a mechanism of solution which is totally different from that of ordinary solutions.

However, if the temperature is lower than the melting point of hydrated solid surfactant, it aggregates infinitely and the solubility is very small. In the case of nonionic surfactants, the hydrophilic portion has to be sufficiently large for it to disperse, forming micelles. At a higher temperature, the hydrophilic property of the surfactant diminishes, and it aggregates infinitely, i.e., a surfactant phase separates above the cloud point (see Chap. 9).

3. *Solution of water-soluble polymers or biocolloids.* Infinite dissolution of water into water-soluble polymers and the

pseudo-phase inversion in solution is considered to be reasonable. This has been discussed in detail in Chap. 8.

4. *Solubilization.* This is a technique to render a water-insoluble substance soluble by adding a surfactant and causing the desired solute to dissolve in the surfactant aggregates, or micelles. Solubilization is primarily applied to the solution of oily or lipophilic substances. For details, see Chap. 10.

REFERENCES

1. H. Hiraska and J. H. Hildebrand, J. Phys. Chem., *67*, 916 (1963); K. Shinoda and J. H. Hildebrand, J. Phys. Chem., *69*, 605 (1965).
2. J. H. Hildebrand, E. T. Ellefson, and C. W. Beebe, J. Am. Chem. Soc., *39*, 2301 (1917).
3. A. G. Leiga and J. N. Sarmousakis, J. Phys. Chem., *70*, 3544 (1966).

EXERCISE

11.1 The molar volume of a solid (component 2) is smaller than its partial molar volume in saturated solution. How will the solubility change if the pressure is increased?

Answer: $V_2^s < \bar{V}_2$. At the equilibrium state,

$$\Delta\bar{G} = \bar{G}_2 - G_S = 0$$

$$\left(\frac{\partial \Delta\bar{G}}{\partial P}\right)_{T,n_1,n_2} = \left(\frac{\partial \bar{G}_2}{\partial P}\right)_{T,n_1,n_2} = \left(\frac{\partial G_2^{s}}{\partial P}\right)_T$$

Here,

$$\left(\frac{\partial \bar{G}_2}{\partial P}\right)_{T,n_1,n_2} = \bar{V}_2, \qquad \left(\frac{\partial G_S}{\partial P}\right)_T = V_2^s$$

Therefore,

$$\left(\frac{\partial \Delta\bar{G}}{\partial P}\right)_{T,n_1,n_2} = \bar{V}_2 - V_2^s > 0$$

Consequently, $\Delta\bar{G}$ increases as the pressure increases. Thus, the equilibrium shifts in the direction of solid separation from the solution state, and the solubility decreases.

Generally, in the case of nonelectrolyte solution, the partial molar volume is larger than the molar volume (liquid) and the molar volume (solid) is even smaller.

Quantitatively, the solubility x_2 changes in accordance with

$$\left(\frac{\partial \ln x_2}{P}\right)_T = \frac{V_2 - \bar{V}_2}{RT}$$

$$RT \ln \frac{x_2 \ (P = P)}{x_2 \ (P = 1)} = (V_2 - \bar{V}_2)(P - 1)$$

APPENDIX A

LIST OF SYMBOLS*

Thermodynamic Quantities[†]

B (B')	coefficient of enthalpy of mixing (per unit volume)
C	heat capacity, concentration of solution
D	permittivity, dielectric constant
E	energy
F	Helmholtz energy • electric field
G	Gibbs energy
H	enthalpy, heat content
K	equilibrium constant
M	molecular weight • molality
N	number of molecules
N_0	Avogadro's number, 6.02×10^{23} molecules/mol
P	pressure
R	gas constant per mole
S	entropy
T	absolute temperature
V	volume

*Minor symbols which occur at only one or two isolated places in the text are not listed.

[†]Partial molal thermodynamic quantities are indicated by an overbar, e.g., $\bar{C}$, $\bar{E}$, $\bar{F}$, $\bar{H}$, $\bar{S}$, $\bar{V}$.

Y	extensive thermodynamic function

Other Symbols

a	van der Waals attraction constant, activity
b	van der Waals repulsion constant
d	distance
f	fugacity
k	Boltzmann constant, gas constant per molecule
ln	natural logarithm
log	common logarithm
n	number of moles
p	vapor pressure, partial pressure
r	radius
t	Celsius temperature
u	potential energy of a pair of molecules
x	mole fraction

Greek

α	coefficient of thermal expansion, polarizability
β	coefficient of compressibility
γ	activity coefficient, surface tension
δ	solubility parameter
Δ	increment, a difference function
μ	chemical potential, dipole moment
$\bar{\mu}$	electrochemical potential
π	3.14159..., osmotic pressure
σ	collision diameter distance parameter, surface charge density
ϕ	volume fraction
ψ	wave function, electrical potential

Superscripts (Processes)

E	excess
F	fusion

M	mixing
S	sublimation
v	vaporization

Superscripts (State)

g	gas
s	solid
o	standard (pure liquid)
ℓ	liquid

Subscripts

b	normal boiling point
c	critical point
m	mixture
p, P	constant pressure
r	reduced
u, v	constant volume
1, 2, 3	components

APPENDIX B

Energy Conversion Factors

	cal/mol	J/mol	cm^3 atm/mol	cm^{-1}	eV	erg/molecule
1 cal/mol	1	4.1840	41.292	0.34974	4.3361×10^{-5}	6.9465×10^{-17}
1 J/mol	0.23901	1	9.8692	0.08359	1.0364×10^{-5}	1.6602×10^{-17}
1 cm^3 atm/mol	0.024218	0.10133	1	8.470 10^{-3}	1.0501×10^{-6}	1.6823×10
1 cm^{-1}	2.8593	11.963	118.07	1	1.2398×10^{-4}	1.9862×10^{-16}
1 eV	23,062	96,493	9.523×10^5	8065.7	1	1.6020×10^{-12}
1 erg/molecule	1.4396×10^{16}	6.023×10^{16}	5.944×10^{17}	5.0348×10^{15}	6.2422×10^{11}	1

ANSWERS TO PROBLEMS

1.1 See Table 3.2.

1.2 $\bar{V}_{H_2O} = 16.25\ \text{cm}^3$

2.1 $-R \ln x_2$

2.2 0.342

3.2 29%

3.3 2.3 atm

3.4 127.4

4.1 $\Delta\bar{E}_2 = \left(\frac{\partial \Delta E_M}{\partial n_2}\right)_{n_1} = V_2\phi_1^{\,2}(\delta_1 - \delta_2)^2$

4.2 1.925, 0.662

4.3 8.6

5.1 $RT_c = \frac{2x_1x_2v_1^{\,2}v_2^{\,2}}{(x_1v_1 + x_2v_2)^3}(\delta_1 - \delta_2)^2,\ x_{1c} = \frac{(v_1^{\,2} - v_1v_2 + v_2^{\,2})^{1/2} - v_1}{v_2 - v_1}$

5.2 11.95 cal/deg mol, 24.35 cal/deg mol

6.1 1, 0.36, 0, 1

8.1 $\Delta\bar{S}_2 = -R\left[\ln \phi_2 + \phi_1\left(1 - \frac{V_2}{V_1}\right)\right]$

8.2 $RT_c = \dfrac{2V_1V_2(\delta_1 - \delta_2)^2}{(V_1^{1/2} + V_2^{1/2})^2}, \qquad \dfrac{x_{2c}}{x_{1c}}\left(\dfrac{V_1}{V_2}\right)^{3/2}$

INDEX